STUDY

for

HEIN, BEST, and PATTISON'S

COLLEGE CHEMISTRY:

AN INTRODUCTION TO GENERAL, ORGANIC, AND BIOCHEMISTRY
Fourth Edition

STUDY GUIDE

for

HEIN, BEST, and PATTISON'S

COLLEGE CHEMISTRY:

AN INTRODUCTION TO GENERAL, ORGANIC, AND BIOCHEMISTRY

Fourth Edition

PETER C. SCOTT
Linn-Benton Community College
Albany, Oregon

Brooks/Cole Publishing Company
Pacific Grove, California

© 1988, 1984, 1980, 1976 by Wadsworth, Inc., Belmont, California 94002. All rights reserved. No part of this book may be reproduced, stored in a retrieval system, or transcribed, in any form or by any means—electronic, mechanical, photocopying, recording, or otherwise—without the prior written permission of the publisher, Brooks/Cole Publishing Company, Pacific Grove, California 93950, a division of Wadsworth, Inc.

Printed in the United States of America

10   9   8   7   6   5   4   3   2   1

ISBN 0-534-08565-2

Sponsoring Editor: Harvey Pantzis
Production Coordinator: Linda Loba
Interior Design: Project Publishing & Design, Inc.
Cover Design: Katherine Minerva
Cover Photo: Courtesy of Nelson Max, Lawrence Livermore Laboratory
Typesetting: Omegatype Typography, Inc.
Printing and Binding: Malloy Lithographing, Inc.

Dedicated to my dear wife, Jeanette

Preface

This study guide has been prepared to accompany *College Chemistry: An Introduction to General, Organic, and Biochemistry, Fourth Edition*, by Hein, Best, and Pattison. The guide selects certain key concepts from the text, and provides the student with a means of self-evaluation for determining how well he or she understands the material of each chapter.

By presenting a slightly different viewpoint and emphasis, the study guide provides students with an approach other than that of the textbook toward mastery of the subject matter. Because this guide is an auxiliary student-oriented aid to complement the textbook, no new material is presented.

The chapters in the guide are organized as follows: The chapter headings are the same as in the text. The self-evaluation section provides various exercises that allow the student to check his or her understanding of the text chapter objectives. Students are asked to complete fill-in responses or to choose an appropriate response. Tables to be constructed and problems to be solved are presented. Nomenclature, equation balancing, and use of the periodic table are covered. The last problems in the self-evaluation section are identified as challenge problems. These will usually be a little more difficult or complex than the other exercises in the chapter. Additionally, the explanations about the answers will be briefer and will assume that the student has a good grasp of the basic material covered up to that particular point. Complete answers to questions and solutions to problems are given at the end of each chapter, to provide immediate reinforcement or correction of errors. The last section in each chapter provides a recap, or summary.

In addition to the usual exercises, crossword puzzles are included in the guide to give the student a break and to check his or her vocabulary skills. The student will probably need a periodic table to help in solving the puzzles.

The author visualizes using the study guide as follows: The student would (1) read the assigned text chapter; (2) go through the self-evaluation process in the guide; (3) return to the assigned homework problems, lab experiments, and other instructor-generated activities; and (4) be tested on predetermined performance objectives for each chapter.

To the Student

This guide, which accompanies the textbook *College Chemistry: An Introduction to General, Organic, and Biochemistry, Fourth Edition*, by Hein, Best, and Pattison, is a student-oriented self-study guide. It has been prepared to help you evaluate your grasp of certain concepts in the textbook; you can then identify trouble spots and ask your instructor for individual help.

What is the place of the study guide in the total chemistry program? Your instructor will hand out reading assignments from the textbook, which will be followed by homework questions and problems and laboratory experiments. After classroom discussion and problem sessions, you will be tested on the material. So, when do you use the study guide? A typical student of mine might first read the assigned chapter in the text (with rereading, if necessary), and then turn to the study guide for a self-evaluation of his or her understanding of the key concepts. All of the answers and solutions are given so that any errors can be quickly spotted. If a particular concept or type of problem doesn't make sense, the student can return to the textbook. Then the student is ready to work the homework assignment, knowing that he or she understands the basic concepts in the chapter. The week's lab experiment is performed, followed by the chapter quiz.

Each chapter of the study guide that has questions to answer and problems to solve is organized in the following manner: The "Self-Evaluation Section" provides various types of exercises that enable you to check your understanding of the key objectives. The answers and solutions to the problems are given at the end of the chapter so that you can find out where your difficulty may have occurred. The challenge problems at the end of each chapter are usually a bit more difficult or complex than the rest of the self-evaluation exercises. You should work on these after you feel comfortable with the other material. The answers to the challenge problems may not be as thorough as the other answers since you're expected to have a good grasp of the basic material before attempting the challenge problems. The last section gives a summary from the chapter and indicates where you are going next in the course.

The study guide allows you to work on your own and make efficient use of your time. It should assist you in finding out whether you have learned the basic concepts.

Best wishes for success in your study of chemistry.

Contents

1 Introduction 1
2 Standards for Measurement 5
3 Properties of Matter 21
4 Elements and Compounds 27
5 Atomic Theory and Structure 37
6 The Periodic Arrangement of the Elements 45
7 Chemical Bonds—
   The Formation of Compounds from Atoms 51
8 Nomenclature of Inorganic Compounds 63
9 Quantitative Composition of Compounds 71
10 Chemical Equations 85
11 Calculations from Chemical Equations 93
12 The Gaseous State of Matter 103
13 Water and the Properties of Liquids 117
14 Solutions 125
15 Ionization. Acids, Bases, Salts 151
16 Chemical Equilibrium 163
17 Oxidation-Reduction 183
18 Radioactivity and Nuclear Chemistry 197
19 Chemistry of Some Selected Elements 203
20 Organic Chemistry, Saturated Hydrocarbons 213
21 Unsaturated Hydrocarbons—
   Alkenes, Alkynes, and Aromatic Hydrocarbons 221
22 Alcohols, Phenols, and Ethers 229
23 Aldehydes and Ketones 237
24 Carboxylic Acids, Their Derivatives, and Amines 245
25 Polymers-Macromolecules 257
26 Stereoisomerism 263

27 Carbohydrates 271
28 Lipids 281
29 Amino Acids, Polypeptides, and Proteins 285
30 Nucleic Acids and Heredity 291
31 Nutrition and Digestion 295
32 Bioenergetics 299
33 Carbohydrate Metabolism 303
34 Metabolism of Lipids and Proteins 307

CHAPTER ONE
Introduction

As a good way to check your readiness to begin a course in chemistry, work through the following mathematics survey. We use this survey for all of our chemistry students, and they have found it helpful in determining areas to be reviewed.

Circle the response for each of the questions that most closely expresses your feelings, as follows:

(a) The problem is worked correctly.
(b) The problem has been worked incorrectly.
(c) I have studied this type of problem before but can't remember how to work it now. I'm not sure if it's right or not.
(d) I have never studied this type of problem before.

Do your calculations in this space or on scratch paper.

1. $2/4 = 0.33$ a b c d
2. $1/16 = 0.0625$ a b c d
3. $1/10 = 0.01$ a b c d
4. $30/6 = 0.5$ a b c d
5. $1/3 \times 4/5 = 5/8$ a b c d
6. $1/2 \div 1/4 = 2$ a b c d
7. $3/4 + 3/5 = 3/20$ a b c d
8. $1/3 - 1/4 = 1/12$ a b c d
9. $\dfrac{2 \times 4 \times 3}{6 \times 2} = 2$ a b c d
10. $10^2 = 100$ a b c d
11. $\sqrt{25} = 5$ a b c d
12. $2^3 = 6$ a b c d
13. $\sqrt[3]{8} = 2$ a b c d

Do your calculations in this space.

14. $6^2 \times 6^3 = 6^6$ a b c d

15. $\dfrac{9^4}{9^3} = 9$ a b c d

16. $\dfrac{a^7}{a^{10}} = a^{-3}$ a b c d

17. $6.9 \times 10^3 = 690$ a b c d

18. $0.0054 = 5.4 \times 10^{-3}$ a b c d

19. $(2 \times 10^3)(5 \times 10^2) = 10^6$ a b c d

20. $(5 \times 10^{-3})(3 \times 10^5) = 1.5 \times 10^2$ a b c d

21. 1594.61 rounded off to four figures is 1595   a b c d

22. The reciprocal of 2 is 1/2   a b c d

23.
$$\begin{array}{r} 6.1 \text{ cm} \\ \times\ 3.5 \text{ cm} \\ \hline 305 \\ 183 \\ \hline 21.35 \text{ cm}^2 \end{array}$$
Is this answer written with the correct number of significant figures?   a b c d

24. If $\dfrac{x}{2.5} = \dfrac{3}{7.5}$, then $x = 1$ a b c d

25. Let $T_1 \times N_1 = T_2 \times N_2$
 If $T_1 = 5, N_1 = 3$ and $T_2 = 6$,
 then $N_2 = 2.5$ a b c d

26. If $\dfrac{P_1 V_1}{T_1} = \dfrac{P_2 V_2}{T_2}$ then $T_2 = \dfrac{P_2 V_2}{P_1 V_1 T_1}$ a b c d

27. If $\dfrac{1.5}{x} = \dfrac{3.0}{6.4}$, then $x = 3.5$ a b c d

28. If it takes 1 calorie to heat 1 gram of water 1°C, how many calories will be required to heat 3 grams of water 5°C? Answer: 8 calories   a b c d

Check your answers from the list on page 3. Students who miss as many as 17 will have a difficult time with the material in the text without an intensive review and extra work outside of class. Keep this in mind if you found that many of the problems were unfamiliar to you or you circled an incorrect response. If you missed about half the problems, you should study thoroughly Appendix I — Mathematical Review at the back of the text.

ANSWERS TO MATHEMATICS SURVEY

1. b	8. a	15. a	22. a
2. a	9. a	16. a	23. b
3. b	10. a	17. b	24. a
4. b	11. a	18. a	25. a
5. b	12. b	19. a	26. b
6. a	13. a	20. b	27. b
7. b	14. b	21. a	28. b

RECAP SECTION

The objectives of Chapter 1 are to introduce you to the broad field of science known as chemistry and to develop some of chemistry's historical aspects. We discuss the importance of chemistry in other fields of science such as biology and agriculture, and we describe the two main branches of chemistry—inorganic and organic.

You should also gain some insight into how the scientific method is applied in chemistry. As you study chemistry or any experimental science, you find the steps of hypothesis, theory, and scientific law leading time and again toward an understanding of natural phenomena.

The importance of problem solving to the science of chemistry cannot be overemphasized. You will be working problems within the text material, solving homework problems, and using this study guide to evaluate your ability to solve certain specific types of problems. Many of the laboratory experiments also involve problem solving.

The techniques you will use are mostly simple arithmetic operations, but you must have a complete understanding of them. Significant figures and scientific notation are also important for your progress in chemistry.

The topics of scientific notation, significant figures and rounding-off numbers are covered thoroughly in Chapter 2 of the text. Remember that additional helpful information which will be of use throughout your study of chemistry is found in Appendix I of the text.

CHAPTER TWO
Standards for Measurement

SELF-EVALUATION SECTION

If you are to talk correctly with your fellow students and instructors about chemical principles and knowledge, you must have an intimate understanding of chemical terminology. Chapter 2 contains some very essential terms, which you must master. Some of the terms will be familiar but will have more precise definitions than you are used to. For instance, though the terms *mass* and *weight* are often used interchangeably, the two words do not have precisely the same meaning. Mass is an inherent property of matter and constitutes an essential characteristic of matter. Weight, on the other hand, is an extraneous property of an object and depends on the force of gravity acting on the object. An object that weighs 100 grams at sea level will weigh less at an altitude of 10,000 feet since the force of gravity is less. If the same object is placed in a space vehicle, it will exhibit weightlessness at some point in space and float around the cabin if not anchored down.

However, if a chemical balance, which is used to measure masses of objects by comparison with known masses, is used to make a mass determination of an object and a value of 225 grams is obtained, then the object's mass is constant no matter where the object is. A determination of the object's mass in a space vehicle would yield the same value, 225 grams, as before. If two objects of equal mass are weighed at the same place on the earth's surface they will have the same weight. The unfortunate circumstance is that there is no simple word in English to describe the operation of making a mass determination, so the verb "to weigh" is used for this important experimental procedure. Keep in mind that when you make a weighing in the chemistry laboratory you are in fact measuring the mass of the object being weighed.

Two other terms introduced in Chapter 2 that are used in everyday conversation are *heat* and *temperature*. There is some confusion about the proper usage of *temperature* and *heat*. It is usual to state that the temperature of an object indicates how much heat the object contains. But this is an incorrect notion. Think of two objects of different size but at the same temperature. Which object contains more heat energy? The larger object, of course. Thus, temperature is only an indication of the degree of "hotness" or "coldness" of an object. The temperature scales are relative, arbitrary scales, established with certain reference points. The Celsius scale, for instance, has 0 degrees set at the freezing point of water and 100 degrees set at the boiling point of water.

Heat is one form of energy just as light and electricity are. When small particles are vibrating, heat energy is given off. Think of the conversion of electrical energy into heat and light energy as an electric stove burner becomes red-hot. It is not too difficult to imagine the intense motion in the heating element as the current of electricity passes through. When we talk about the amount of heat energy involved in a chemical process, we use the SI unit of measurement called the *joule*. We define the joule by an experimental procedure—a common practice in science. If we measure out 1 gram of water and place it at 14.5° Celsius, then 4.184 joules of heat energy from a match, gas burner, or other source will raise the temperature of the 1 gram of water 1° Celsius to 15.5° Celsius. In words, the number of joules gained or lost is equal to the mass of water times the temperature change times a constant. The constant is the *specific heat* for the material in question. For water, the specific heat is 4.184 J/g°C. You will work some sample problems later.

The last term defined in this section is *density*. Again, we can use either an algebraic form of definition or the English equivalent. The density of a substance is related to its mass and volume. Since all matter in the universe has mass and occupies space or volume, the quantity defined as density is a useful physical characteristic used in describing various substances. The defining equation for density says that the density of a substance is equal to the mass divided by the volume. The mass is given in grams, and the volume can be expressed in cubic centimeters, milliliters, or liters. The units for density are therefore either grams/cubic centimeter (g/cm^3), grams/milliliter (g/mL), or grams/liter (g/L).

We have just finished discussing a few of the important new terms used in chemistry. Did you notice that the definitions of *joule* and *density* each involved a simple algebraic equation with units? Thus, density is defined as the mass per unit volume (d = g/mL, for example). No numbers or measurements are involved in these definitions, only units. After completing a review of the metric system, you will work through sample problems involving heat changes and the densities of substances using the factor-label or dimensional analysis technique. The factor-label approach will be used as consistently as possible throughout.

1. In order to handle large numbers efficiently, it is worthwhile to use scientific notation, which means writing a number as a power of 10. The numbers are always written between 1 and 10 with the associated power of 10.

    (1) Express the following numbers as powers of 10.

    2510 _____

    0.0065 _____

    46 _____

    77,000,000 _____

    0.0102 _____

    (2) Express the following in decimal form.

    4.77×10^4 _____

8.41×10^{-2} _____

5.8×10^{1} _____

9.1×10^{0} _____

1.415×10^{2} _____

2. Review of metric units, abbreviations, and definitions. Fill in the correct responses.

In our study of chemistry, we will be using the metric system of measurement routinely. On any one occasion, we may have the task of making measurements of one of several quantities, which might include (1) _____, (2) _____, or (3) _____. For the measurement of mass in the metric system, we will use the unit named (4) _____ with the abbreviation (5) _____. If we are measuring the length of an object, as in the case of certain density experiments, we will use the unit called (6) _____, which is abbreviated (7) _____. However, probably the most common type of measurement we make in a chemistry laboratory involves the metric unit of volume (8) _____, which is abbreviated (9) _____.

To these three units of measure for mass, length, and volume, we can attach prefixes to change the unit's values by various powers of 10. For example, *kilo* means (10) _____ and *milli* means (11) _____. The prefix for 0.01 (10^{-2}) is (12) _____, and the prefix for 0.000001 (10^{-6}) is (13) _____. The abbreviations for the prefixes *kilo, centi, micro,* and *milli* are (14) _____, (15) _____, (16) _____, and (17) _____.

Using the four prefixes in general use in chemistry, we must be able to convert from one metric unit to any other corresponding metric unit. We will use a series of conversions to find out if a series of decimal-point changes presents any problems to you.

Convert These Units **To These Units**

(18) 42 g _____ mg

(19) 0.741 cm _____ m

(20) 8.4 mL _____ L

(21) 776 kg _____ g

(22) 1005μL _____ mL

Convert These Units	To These Units
(23) 15μg	_____ mg
(24) 6.2 L	_____ mL
(25) 0.34 km	_____ cm
(26) 1.25 mg	_____ g
(27) 0.0089 g	_____ mg
(28) 9 cm	_____ μm

3. Express the following numbers in exponential form

   (1) 67,000 _____ (4) 0.00078 _____

   (2) 0.0654 _____ (5) 411,000 _____

   (3) 10,000,000 _____ (6) 10 _____

4. Express the following numbers in decimal form.

   (1) 4.8×10^3 _____ (4) 1074×10^{-2} _____

   (2) 0.67×10^2 _____ (5) 0.0034×10^3 _____

   (3) 0.151×10^{-5} _____ (6) 11×10^{-5} _____

5. Round off the following numbers to four digits.

   (1) 47.679 _____ (5) 0.077938 _____

   (2) 1.0255 _____ (6) 21.6251 _____

   (3) 100.2484 _____ (7) 5.00515 _____

   (4) 16.0502 _____ (8) 74.2856 _____

6. Solve for x.

   (1) $x = (1.3 \times 10^2)(6.4 \times 10^{-1})$

   (2) $x = (5.76 \times 10^{-4})(2.4 \times 10^3)$

   (3) $(8.9 \times 10^1)x = (5.46 \times 10^3)$

   (4) $(0.77 \times 10^4)x = 125{,}000$

7. Converting American units to metric units.
   (1) A college basketball player playing forward is generally about 6 ft 6 in. tall. How many centimeters is this? How many meters?

   Do your calculations here.

(2) Tall cans of juice contain 46.0 fluid ounces. How many mL, is this? One ounce equals 29.6 mL.

Do your calculations here.

(3) A can of green beans cost 59¢ and weighs 1 pound. What is the cost in cents per gram? The words "cents per gram" are a way of expressing the fraction cents/gram. This means that the number of cents is divided by the number of grams. The word "per" is a common method used in science to indicate a division operation or a fraction. Be alert to this term in future problems.

1 lb = 454 g

Do your calculations here.

8. We can convert the Kelvin temperature scale to the Celsius scale, and vice versa, with the formula K = °C + 273. The value 273 is a constant value. Using the formula, make the following conversions.

Convert From	To
(1) 25°C	_____ K
(2) 273K	_____ °C
(3) −15°C	_____ K
(4) −273°C	_____ K
(5) 398K	_____ °C
(6) 100°C	_____ K
(7) 58K	_____ °C

Using the equations given for converting °F to °C, and vice versa, make the conversions asked for.

$$°F = (1.8 \times °C) + 32$$

$$°C = \frac{(°F - 32)}{1.8}$$

(8) It is not uncommon for temperatures in the Canadian plains to reach −60°F and below during the winter. What is this temperature in °C?

Do your calculations here.

(9) Sodium chloride melts at 801°C. What is this temperature in °F?

Do your calculations here.

9. We now return to the problems concerning heat, joules, and temperature changes in samples of water.

Remembering that the number of joules (abbreviated J) gained or lost during heat energy changes is equal to the mass of water times the specific heat of the substance times the temperature difference, we can write the following equation.

$$\text{joules} = (\text{grams of water})(\text{specific heat of substance})(\Delta t)$$

The symbol Δt means "temperature difference." Temperature difference is found by subtracting the initial temperatures of the water from the final temperature. It is important to do the subtraction in this order. The last term in the equation is the specific heat for the substance in question. For water, the specific heat is 4.184 J/g°C. An equation with units attached would be

$$\text{joules} = g \times 4.184 J/g°C \times °C$$

Try the following problems.
(1) How many joules of heat energy are required to heat 330 g of water from 25°C to boiling (100°C)?

Do your calculations here.

(2) How many grams of water can be heated from 5°C to 52°C with 1000 J of heat energy?

Do your calculations here.

(3) Let's work problems (1) and (2) again but this time we will use the calorie unit of heat measurement. Recall that the specific heat of water in calorie units is 1 cal/g°C or in words "one calorie of heat energy will raise one gram of water one degree Celsius." Therefore, how many calories of heat energy are required to heat 330g of water from 25°C to boiling (100°C)?

Do your calculations here.

(4) In a similar manner, how many grams of water can be heated from 5°C to 52°C with 1000 cal of heat energy?

Do your calculations here.

10. As you progress in your study of various substances, you will find a great diversity in the physical nature of matter. There are many ways to categorize matter in trying to relate the properties of one substance to another. One such way is through a relationship called *density*. The density of a substance is not only related to its mass but also to the volume it occupies. Thus, a cube of aluminum metal measuring 10 cm on a side weighs much less than a cube of iron metal of the same volume. We say the density of aluminum is less than that of iron.

The equation defining density is written as follows:

$$\text{density} = \frac{\text{mass}}{\text{volume}}$$

$$d = \frac{g}{cm^3} = \frac{g}{mL} \text{ or } \frac{g}{liter}$$

The units for density vary depending on the units used for volume.

(1) Sample problem: Our block of aluminum, 10 cm on a side, was found to have a mass of 2700 g by using a chemical balance. What is the density of aluminum?

Do your calculations here.

(2) Another problem: Logs of black ironwood, which do not float in water, are found to have a density of 1.077 g/cm^3. What is the volume of a piece of ironwood that weighs 750 g?

Do your calculations here.

Challenge Problems

11. The standard of metric length is the meter which was defined (prior to 1983) as 1,650,763.73 wavelengths of a spectral line of the element krypton. To three significant figures determine this wavelength in nanometers if $1 \text{ nm} = 10^{-9} \text{ m}$.

   Do your calculations here.

12. A 100 g piece of copper at 212°F is dropped into 100 g of ethyl alcohol at 72°F. The specific heat of copper is $0.0921 \frac{\text{cal}}{\text{g°C}}$, that of ethyl alcohol is $0.511 \frac{\text{cal}}{\text{g°C}}$. Calculate the final temperature of the mixture.

   Do your calculations here.

13. For each of the following, identify (a) what is the numerical value and what is the unit, (b) how many significant figures, (c) which numbers are known and which are estimated. Round each number to two significant figures (d), and (e) express that number as a power of 10.

   (1) 1055 Kg

   (2) 1,650,763 wavelengths

   (3) 1.077 g/cm^3

(4) 0.00845 L

(5) 776,000 g

RECAP SECTION

This has been a long chapter, but if you feel comfortable with the new terms and techniques, you've come a long way toward being successful in the remaining weeks of the course. We have reviewed some basic terms and concepts, brushed up on the metric system, and practiced converting one metric unit to another. The temperature scales of the most concern are the Celsius and Kelvin scales. The conversion between the two scales is very simple since the size of the degree is the same. Perhaps the most important aspect of the chapter was putting your algebraic skills to use in various practical ways to solve chemical problems. You will find that most chemical problems can be handled by a few simple algebraic techniques. If you were successful with the problems in Chapter 2, you are ready to proceed. If you had difficulty, you should go back over the Math Review in the text and then return to the study guide.

ANSWERS TO QUESTIONS AND SOLUTIONS TO PROBLEMS

1. (1) 2.51×10^3; 6.5×10^{-3}; 4.6×10^1; 7.7×10^7; 1.02×10^{-2}
   (2) 47,700; 0.0841; 58; 9.1; 141.5

2. (1) (2) (3) mass, length, volume, density, or temperature are possible answers
   (4) kilogram   (5) kg   (6) meter   (7) m   (8) liter
   (9) 1   (10) 1000   (11) 0.001   (12) centi- (13) micro-
   (14) k   (15) c   (16) μ (17) m

(18) 4.2×10^4 mg 　　$42 \, g \times 1000 \, \dfrac{mg}{1 \, g} = 42{,}000 \text{ mg} = 4.2 \times 10^4 \text{ mg}$

(19) 0.00741 m 　　$0.741 \, cm \times \dfrac{1 \text{ m}}{100 \, cm} = 0.00741 \text{ m}$

(20) 0.0084 L 　　$8.4 \, mL \times \dfrac{1 \text{ liter}}{1000 \, mL} = 0.0084 \text{ L}$

(21) 7.76×10^5 g 　　$776 \, kg \times \dfrac{1000 \text{ g}}{1 \, kg} = 776{,}000 \text{ g} = 7.76 \times 10^5 \text{ g}$

(22) 1.005 mL 　　$1005 \, \mu L \times \dfrac{1 \text{ mL}}{1000 \, \mu L} = 1.005 \text{ mL}$

(23) 0.015 mg 　　$15 \, \mu g \times \dfrac{1 \text{ mg}}{1000 \, \mu g} = 0.015 \text{ mg}$

(24) 6.2×10^3 mL $6.2 \text{ liter} \times \dfrac{1000 \text{ mL}}{1 \text{ liter}} = 6200 \text{ mL} = 6.2 \times 10^3$ mL

(25) 3.4×10^4 cm $0.34 \text{ km} \times \dfrac{1000 \text{ m}}{1 \text{ km}} \times \dfrac{100 \text{ cm}}{1 \text{ m}} = 34{,}000 \text{ cm} = 3.4 \times 10^4$ cm

(26) 0.00125 g $1.25 \text{ mg} \times \dfrac{1 \text{ g}}{1000 \text{ mg}} = 0.00125$ g

(27) 8.9 mg $0.0089 \text{ g} \times \dfrac{1000 \text{ mg}}{1 \text{ g}} = 8.9$ mg

(28) 9×10^4 μm $9 \text{ cm} \times \dfrac{1 \text{ m}}{100 \text{ cm}} \times \dfrac{10^6 \text{ μm}}{\text{m}} = 9 \times 10^4$ μm

3. (1) 6.7×10^4 (4) 7.8×10^{-4}
   (2) 6.54×10^{-2} (5) 4.11×10^5
   (3) 1×10^7 (6) 1.0×10^1

4. (1) 4800   (4) 10.74
   (2) 67   (5) 3.4
   (3) 0.00000151   (6) 0.00011

5. (1) 47.68   (4) 16.05   (7) 5.005
   (2) 1.026   (5) 0.07794   (8) 74.29
   (3) 100.2   (6) 21.63

6. (1) $x = 83$
   (2) $x = 1.4$
   (3) $x = 61$
   (4) $x = 16$

7. (1) We need to change 6 ft 6 in. into inches and then convert to the metric units of centimeters and meters.

$$6 \text{ ft } 6 \text{ in.} = 6 \text{ ft} \times \dfrac{12 \text{ in.}}{\text{ft}} + 6 \text{ in.} = 78 \text{ in.}$$

$$78 \text{ in.} \times \dfrac{2.54 \text{ cm}}{\text{in.}} = 2.0 \times 10^2 \text{ cm}$$

$$2.0 \times 10^2 \text{ cm} \times \dfrac{1 \text{ m}}{100 \text{ cm}} = 2.0 \text{ m (2 sig. figures)}$$

(2) Since 1 oz. = 29.6 mL we need to set up the problem to cancel out oz. and give an answer with units of mL.

$$46.0 \text{ oz.} \times \dfrac{29.6 \text{ mL}}{1 \text{ oz.}} = 1.36 \times 10^3 \text{ mL (3 sig. figures)}$$

(3) 13¢ per g
We must first convert 1 pound to grams and then divide the cost by the number of grams.

$$1 \text{ lb} \times \frac{454 \text{ g}}{\text{lb}} = 454 \text{ g}$$

59¢/454g = 13¢ per g (2 sig. figures)

8. (1) 298K   (2) 0°C   (3) 258K   (4) 0K
(5) 125°C   (6) 373K   (7) −215°C

(8) $°C = \dfrac{-60° - 32}{1.8} = \dfrac{-92°}{1.8} = -51°C$

(9) °F = (1.8 × 801°) + 32 = 1470°F (3 sig. figures)

9. (1) 100,000 joules or 1.0×10^5 joules to 2 sig. figures. We know the mass of water (330g) and can determine the temperature difference by subtraction. The unknown quantity is the number of joules of heat energy required. Make the necessary substitutions in the equation.

   joules = 330g × 4.184J/g °C × (100°C − 25°C)

   = 330g × 4.184J/g °C × 75°C

   = 103,554

   = 1.0×10^5 joules to 2 significant figures

(2) 5.1 g of water (2 sig. figures)
We are asked to determine the mass of water which 1000 joules of heat energy can raise from 5°C to 52°C.
Solving our equation for grams we have

   joules = g × specific heat × Δt

   $g = \dfrac{\text{joules}}{\text{specific heat} \times \Delta t}$

   $= \dfrac{1000 \text{ joules}}{4.184 \text{J/g°C} \times (52°C - 5°C)}$

   $= \dfrac{1000 \text{ joules}}{4.184 \text{J/g°C} \times 47°C}$

   = 5.1 g of water

(3) 25,000 cal to 2 sig. figures
We know the mass of water (330 g) and know the temperature difference by subtraction. The unknown quantity again is the number of calories of heat energy required. Make the necessary substitutions in the equation.

   cal = 330 g × 1 cal/g°C × (100°C − 25°C)

   = 330 g × 1 cal/g°C × 75°C

   = 24,750 cal

   = 2.5×10^4 cal (2 sig. figures)

(4) 21 g of water

Again, what mass of water can 1000 cal raise from 5°C to 52°C.

Solving our equation for grams we will have

$$\text{cal} = g \times \text{specific heat} \times \Delta t$$

$$g = \frac{\text{cal}}{\text{specific heat} \times \Delta t}$$

$$= \frac{1000 \text{ cal}}{1 \text{ cal/g}°C \times (52°C - 5°C)}$$

$$= \frac{1000 \cancel{\text{cal}}}{1 \cancel{\text{cal/g}°C} \times 47\cancel{°C}}$$

= 21 g of water

10. (1) 2.7 g/cm³

The mass is given as 2700 g, and we need to determine the volume of a cube 10 cm on a side. Volume is length × width × height. Multiplying 10 cm × 10 cm × 10 cm equals 1000 cm³. Substituting into our equation, we have

$$d = \frac{\text{mass}}{\text{volume}} = \frac{2700 \text{ g}}{1000 \text{ cm}^3} = 2.7 \frac{\text{g}}{\text{cm}^3}$$

The answer says that aluminum metal weighs 2.7 g per each cm³. Therefore, 2 cm³ would weigh 5.4 g, and so forth.

(2) 696 cm³

We are given the density of the wood and the mass and asked to solve for volume. We write down the equation, then isolate the unknown quantity and take a close look at how the units of a problem can help us in solving equations.

$$d = \frac{\text{mass}}{\text{volume}}$$

Multiply each side by volume.

$$\text{volume} \times d = \frac{\text{mass} \times \cancel{\text{volume}}}{\cancel{\text{volume}}}$$

Now divide each side by density.

$$\frac{\text{volume} \times \cancel{d}}{\cancel{d}} = \frac{\text{mass}}{d}$$

Substitute our values into the modified equation.

$$\text{volume} = 750 \cancel{\text{g}} \times \frac{1 \text{ cm}^3}{1.077 \cancel{\text{g}}}$$

= 696 cm³ (3 sig. figures)

Notice that "g" cancels out then leaves our answer with the units for volume, which are the desired units.

11. From the information given, if 1 nm = 10^{-9} m, the 1 m = 10^9 nm. Therefore, we can say that

$$1 \text{ m} = 10^9 \text{nm} = 1{,}650{,}763.73 \text{ wavelengths}$$

To find the length of one wavelength we can set up the following two relationships:

$$10^9 \text{nm} = 1{,}650{,}763.73 \text{ wavelengths}$$

$$\frac{10^9 \text{nm}}{1{,}650{,}763.73} = 1 \text{ wavelength (dividing each side by } 1{,}650{,}763.73)$$

$$606 \text{ nm} = 1 \text{ wavelength (to three significant figures)}$$

12. In examining the problem we should notice first that the temperatures are given in °F while specific heats contain the temperature unit of °C. In general, as we work with chemical problems we will find that conversions are necessary when we see temperatures given as °F. Therefore, the copper metal's temperature is

$$°C = \frac{(°F - 32)}{1.8} = \frac{(212 - 32)}{1.8} = 100°C$$

The temperature of the ethyl alcohol is

$$°C = \frac{(°F - 32)}{1.8} = \frac{(72 - 32)}{1.8} = 22°C$$

Next we can say that the copper metal piece has been heated up, dropped into a cooler material, the alcohol, and caused the final mixture to increase in temperature. The calories of heat gained by the copper as it was heated are lost to the alcohol so that the final temperature of the copper and alcohol are equal.

In order to find the final temperature, we should first calculate the number of calories of heat energy available in the heated piece of copper.

$$\text{grams} \times \text{(specific heat)} \times °C = \text{calories}$$

$$100 \text{ g} \times \left(0.0921 \frac{\text{cal}}{\text{g} °C}\right) \times 100 °C = \text{calories}$$

$$= 921 \text{ cal}$$

This amount of heat energy is available to raise the temperature of the alcohol to some final value. Using the specific heat of alcohol and the information given we can write the following equation:

$$\text{(grams of alcohol)} \times \text{(specific heat of alcohol)} \times \Delta T = 921 \text{ calories}$$

$$(100 \text{ g}) \times \frac{0.511}{\text{g}°C} \times (t_{\text{final}} - t_{\text{initial}}) = 921 \text{ calories}$$

$$100 \text{ g} \times 0.511 \text{ cal/g}°C \times (t_{\text{final}} - 22°C) = 921 \text{ calories}$$

Canceling and rearranging gives us

(51.1) $(t_{final}) = 921 + 1124$

$$t_{final} = \frac{2045}{51.1}$$

$= 40°C$ (to 2 sig. figures)

13. (1) 1055 Kg: (a) 1055 is numerical value, Kg is the unit (b) 4 sig. figures (c) 105 are known, last 5 estimated (d) 1100 Kg to 2 sig. figures (e) 1.1×10^3 Kg

(2) 1,650,763 wavelengths: (a) 1,650,763 is numerical value, wavelengths are the units (b) 7 sig. figures (c) 165076 are known, last 3 is estimated (d) 1,700,000 wavelengths to 2 sig. figures (e) 1.7×10^6 wavelengths.

(3) 1.077 g/cm^3: (a) 1.077 is numerical value, g/cm^3 are the units (b) 4 sig. figures (c) 1.07 known, last 7 estimated (d) 1.100 g/cm^3 to 2 sig. figures (e) 1.1×10^0 g/cm^3

(4) 0.00845 L (a) 0.00845 is numerical value, L is the unit (b) 3 sig. figures (c) 84 known, 5 estimated (d) 0.0085L to 2 sig. figures (e) 8.5×10^{-3} L

(5) 776,000g (a) 776,000 is numerical value, g is unit (b) 3 sig. figures (c) 77 known, 6 estimated (d) 780,000g to 2 sig. figures (e) 7.8×10^5 g

CHAPTER THREE
Properties of Matter

SELF-EVALUATION SECTION

1. All matter in the universe can be classified into one of three states—gas, liquid, or solid. Determine to which of the three states of matter each of the following descriptions relates.

Description	State of Matter
(1) Has a definite fixed shape.	_____
(2) Particles flow over each other while retaining fixed volume.	_____
(3) Exerts a pressure on all walls of the container.	_____
(4) Exhibits no or very slight compressibility.	_____
(5) Particles move independently of each other.	_____
(6) Particles arranged in regular, fixed geometric pattern.	_____
(7) Exhibits slight compressibility.	_____
(8) Exhibits very high compressibility.	_____

2. Statements about the properties and changes of various substances are given below. Place the appropriate letter of identification in each space provided.

   a. physical property   b. chemical property
   c. physical change   d. chemical change

Description	Identification
(1) Potassium is a solid at room temperature.	_____
(2) Potassium is among the most reactive elements in nature.	_____
(3) The melting point of potassium is 63.6°C.	_____

Description	Identification
(4) Potassium reacts vigorously with water.	_____
(5) Potassium added to water produces hydrogen gas plus a water solution of potassium hydroxide.	_____
(6) At 760°C, potassium begins to boil and to change from a liquid to a gas.	_____
(7) At high temperatures, potassium reacts with hydrogen gas to form a metallic hydride.	_____
(8) Alkali metals, such as potassium, form salts with many other elements.	_____
(9) Potassium metal is shiny and fairly soft.	_____
(10) Potassium is useful in the manufacture of fertilizer compounds.	_____

3. Examine the following list of items. Some are mixtures, others are elements or compounds. Place an *S* next to the pure substances and an *M* next to the mixtures.

   (1) wood _____
   (2) sodium chloride _____
   (3) milk _____
   (4) oxygen _____
   (5) rubber _____
   (6) water _____
   (7) air _____
   (8) chlorine _____
   (9) soil _____
   (10) gasoline _____

4. In the equations for the chemical reactions given below, identify which substances are the reactants and which are the products.

   (1) Carbon + Oxygen → Carbon monoxide
   (2) Sodium oxide + Water → Sodium hydroxide
   (3) Nitric acid + Calcium hydroxide → Calcium nitrate + Water
   (4) Aluminum hydroxide + Sulfuric acid → Aluminum sulfate + Water
   (5) Potassium chlorate → Potassium chloride + Oxygen

5. Distinguish between *potential* and *kinetic* energy.

After this section, we will be able to define the two types of energy that matter possesses and to distinguish between them. Energy is simply defined as the capacity to do work. Einstein showed us that mass and energy are directly related by means of his famous equation $E = mc^2$, which says that the energy content of an object equals the mass of the object times the speed of light squared.

Potential energy is the energy which matter has as a result of chemical bonds or because of its position in relationship to another place; kinetic energy is strictly the energy of motion. Filling in the items in this story about a bike rider may help to make the difference between potential energy and kinetic energy clearer to you. Use the symbols *P* (for *potential*) and *K* (for *kinetic*) in the spaces provided to identify the type of energy described by the underlined word or within the statement.

One fine summer morning, as the birds were chirping and flitting (1) _____ _____ about his window, Michael awoke and thought about a bike ride. He scrambled (2) _____ out of bed, dressed, and hurried to breakfast. His breakfast of bacon, eggs, toast, and milk was a source of (3) _____ energy for his ride. He wheeled (4) _____ his bike out of the garage and began converting the (5) _____ energy of his breakfast into (6) _____ energy as he pumped up the hill. As he pedaled, his foot had greater (7) _____ energy at the top of the sprocket than it had at the bottom. As his foot started down, this higher (8) _____ energy was converted into (9) _____ energy. As Michael picked up speed, his bike increased in (10) _____ energy. After pedaling vigorously, he reached the top of the hill and paused. In relationship to the bottom of the hill, he and his bike had increased in (11) _____ energy. As he turned around and started down the hill, he began to convert this (12) _____ energy into (13) _____ energy. All Michael had to do was coast back down and enjoy the ride.

6. Determine whether the following statements or blanks relate to *chemical, light, heat, mechanical,* or *electrical* energy.

    (1) Radiant energy from the sun is used in the process of photosynthesis.   (1) _____
    Photosynthesis involves a conversion of radiant energy into the __(2)__ energy of sugars, starch, and other compounds.   (2) _____
    (3) Organisms utilize the energy of foodstuffs for movement and activity   (3) _____
    (4) and, in the case of warm-blooded animals, for maintaining body temperature.   (4) _____
    When wood is burned, the two forms of energy that are released are __(5)__ and __(6)__ .   (5) _____   (6) _____
    The chemical energy of a car battery is converted into __(7)__ energy, which turns the starter motor.   (7) _____
    Fuels such as gasoline contain __(8)__ energy which, through the process of combustion, is converted to the __(9)__ energy of moving pistons, valves, gears, and wheels.   (8) _____   (9) _____

Hydroelectric dams use the energy of falling water to turn turbines that convert __(10)__ energy into __(11)__ energy.

(10) _____

(11) _____

7. As our last section indicates, the various forms of energy can be converted from one form to another; nothing was said about gaining energy or losing energy during these changes. Various careful experiments have shown that energy cannot be created or destroyed; it can only be transformed from one form to another. The constancy of energy is summarized as the Law of Conservation of Energy. A little earlier in the text, you read that during ordinary chemical reactions the total mass of the reactants was equal to the total mass of the products. Thus, during chemical reactions, mass is neither created nor destroyed. The summarizing statement concerning this experimental fact is called the Law of Conservation of Mass.

If neither energy nor mass can be created or destroyed, might not energy and mass be intimately related? This notion is represented by Einstein's equation. The equation says that the energy of a system is equal to the mass of the system times the speed of light squared. Whenever energy is absorbed or released during a reaction, as when wood burns, there must be a corresponding change in mass. You should now be able to write a statement describing the Law of Conservation of Mass and a statement describing the Law of Conservation of Energy. In addition, since mass and energy are interchangeable, it would be appropriate to state that, in any system, the total mass and energy will be constant regardless of the transformations taking place.

8. You are given a 150 g sample of alumina (Al_2O_3) prepared from the mineral, bauxite. If the percent composition of alumina is known to be 53% Al and 47% O_2, calculate the maximum amount of aluminum that can be obtained from the sample.

Do your calculations here.

Challenge Problem

9. Magnesium ribbon, weighing 1.54 g, is burned in air. The white powdery residue is analyzed and found to contain only magnesium and oxygen. The residue weight is 2.52 g. Calculate the percent composition of the residue, magnesium oxide.

Do your calculations here.

RECAP SECTION

After finishing Chapter 3, you may feel that all the new terms are a bit too much. But the underlying concepts of the unit are simple and the terms just help to define them. The central theme is that matter and energy are interchangeable. Whether matter undergoes a physical change or a chemical change, energy is conserved. Energy takes many forms, just as matter exists in three states. You should be alert to energy forms in your surroundings. It has been said that increasing demands for energy are among our most pressing environmental problems. The material in Chapter 3 allows you to talk about chemistry and your laboratory experiments in a more meaningful way. We are now ready to learn about chemical elements and compounds and how they are put together in an orderly fashion.

ANSWERS TO QUESTIONS AND SOLUTIONS TO PROBLEMS

1. (1) solid   (2) liquid   (3) gas   (4) solid
   (5) gas   (6) solid   (7) liquid   (8) gas

2. (1) a   (2) b   (3) a   (4) b
   (5) d   (6) c   (7) d   (8) b
   (9) a   (10) b

3. (1) M   (2) S   (3) M   (4) S
   (5) M   (6) S   (7) M   (8) S
   (9) M   (10) M

4. (1) Reactants are carbon and oxygen; product is carbon monoxide.
   (2) Reactants are sodium oxide and water; product is sodium hydroxide.
   (3) Reactants are nitric acid and calcium hydroxide; products are calcium nitrate and water.
   (4) Reactants are aluminum hydroxide and sulfuric acid; products are aluminum sulfate and water.
   (5) Reactant is potassium chlorate; products are potassium chloride and oxygen.

5. (1) K   (2) K   (3) P   (4) K
   (5) P   (6) K   (7) P   (8) P
   (9) K   (10) K   (11) P   (12) P
   (13) K

6. (1) light   (2) chemical   (3) mechanical   (4) heat
   (5) heat   (6) light   (7) electrical   (8) chemical
   (9) mechanical   (10) mechanical   (11) electrical

7. No answers necessary.

8. 80 g. The sample weight is given as 150 g of which 53% is aluminum. Therefore

   150 g $\times$ 0.53 (percent expressed as decimal) = 80 g

   The answer on a calculator will be given as 79.5 g but we must use only two significant figures because of the percent composition value.

9. 61% magnesium, 39% oxygen. The total weight of the magnesium oxide which is the reaction product is 2.52 g. The weight of one reactant, magnesium, is given as 1.54 g. The weight of the other reactant, oxygen, can be found by subtraction.

$$2.52 \text{ g} - 1.54 \text{ g} = 0.98 \text{ g O}_2$$

The percent composition for magnesium is found from the following relationship:

$$\% \text{ magnesium} = \frac{\text{weight magnesium}}{\text{weight of magnesium oxide}} \times 100\% = \frac{1.54 \text{ g}}{2.52 \text{ g}} \times 100\% = 61\%$$

likewise

$$\% \text{ oxygen} = \frac{\text{weight oxygen}}{\text{weight of magnesium oxide}} \times 100\% = \frac{0.98 \text{ g}}{2.52 \text{ g}} \times 100\% = 39\%$$

The total percent composition adds up to 100%.

CHAPTER FOUR
Elements and Compounds

SELF-EVALUATION SECTION

1. Fill in with a response appropriate to Chapter 4.

   There are 108 presently known (1) _____, of which at least 88 occur naturally on earth. These building-block units can be combined by means of chemical reactions to form a multitude of chemical (2) _____. Thus, one can see that compounds composed of two or more (3) _____ can be formed or broken down into the constituent (4) _____ by various reactions. When an element is subdivided into smaller and smaller pieces, the final indivisible particle that is left is called an (5) _____. Atoms can combine in two ways in simple, whole-number ratios. Atoms form small, uncharged units called (6) _____. An example is sugar, which can be subdivided into smaller and smaller particles until we finally obtain one (7) _____ of sugar. We cannot divide this unit further without destroying its identity. In other compounds, atoms exist as electrically charged species called (8) _____. These charged species may be composed of one or several atoms. If the ion is positively charged (+), it is called a (9) _____. If the ion is negatively charged (−), it is called a (10) _____. Compounds consisting of ions do not form (11) _____. The formula for an ionic compound represents the smallest whole-number ratio or ions present in the crystalline structure.

   When a chemist analyzes a chemical compound, the percentage composition of the compound by weight is found to be constant. For example, water always contains 11.2% hydrogen and 88.8% oxygen. This experimental observation illustrates the Law of (12) _____ _____. The Law of Definite Composition states that the percentage composition of a pure compound is (13) _____.

2. Write the chemical symbol for the five most abundant elements in the earth's crust, seawater, and atmosphere. Also write the six most abundant elements in the human body. Which elements are common to both lists?

Earth's Chemicals	Human Body
a. _____	a. _____
b. _____	b. _____
c. _____	c. _____
d. _____	d. _____
e. _____	e. _____
	f. _____

Elements common to both:

3. Write the correct chemical symbols for the elements listed in Column A. Likewise, write out the correct chemical names for the symbols listed in Column B. Study Table 4.3 in the text thoroughly before writing this exercise.

Column A		Column B	
Aluminum	_____	Ar	_____
Antimony	_____	Ba	_____
Arsenic	_____	Bi	_____
Boron	_____	Br	_____
Cadmium	_____	Cl	_____
Calcium	_____	Cr	_____
Carbon	_____	Co	_____
Copper	_____	F	_____
Gold	_____	I	_____
Helium	_____	Mn	_____
Iron	_____	Ne	_____
Lead	_____	Ni	_____
Magnesium	_____	P	_____
Mercury	_____	Si	_____
Nitrogen	_____	Na	_____
Potassium	_____	S	_____
Silver	_____	Ti	_____
Tin	_____	U	_____
Tungsten	_____		
Zinc	_____		

4. Look at the following list of chemicals. From what you've learned in Chapter 4, try to identify each formula as to whether it represents a compound or not, and if the formula is that of a compound whether the compound is molecular or ionic in nature as described in Chapter 4.

Formula	Compound? Yes or No	If yes then	Molecular	or	Ionic
(1) I_2	_____		_____		_____
(2) CO	_____		_____		_____
(3) P_4	_____		_____		_____
(4) NaCl	_____		_____		_____
(5) H_2O	_____		_____		_____
(6) CH_4	_____		_____		_____
(7) Fe	_____		_____		_____
(8) Co	_____		_____		_____
(9) He	_____		_____		_____
(10) NH_3	_____		_____		_____

5. Taking each formula in turn from the following list, describe the formula in terms of the number of atoms of each element present in one formula unit of the compound or in multiple numbers of formula units as indicated.

For example, the formula unit of H_2O contains three atoms—2 atoms H and 1 atom O.

(1) H_2SO_4 _____

(2) NaOH _____

(3) $K_2Cr_2O_7$ _____

(4) $Ca(OH)_2$ _____

(5) 2 P_2O_5 _____

(6) $(NH_4)_3PO_4$ _____

(7) $KMnO_4$ _____

(8) 2 $NaHCO_3$ _____

(9) $CuSO_4$ _____

(10) 3 Fe_2O_3 _____

6. Which of the compounds in question 5 are binary compounds? _____

7. Fill in the blank space or circle the appropriate response.

It is often difficult to tell if a material we have obtained in an experiment is a pure substance, such as a compound, or a mixture of compounds. If the characteristics that distinguish a compound from a mixture are understood, sometimes a decision can be made or an experiment performed to tell if we, in fact, have a pure compound. The elemental composition of a compound is fixed in definite proportions by weight as stated in the Law of (1) _____. A mixture may be composed of compounds or elements in (2) varying/fixed proportions. The component elements of a pure chemical compound can be obtained only by (3) physical/chemical changes while a mixture can be separated by (4) physical/chemical means. Identification of the components of a mixture shows that each individual fraction (5) has/has not lost its identity upon forming a mixture while a chemical compound (6) does/does not resemble the component elements.

Indicate by a *C* or an *M* whether the following statements refer to a *compound* or a *mixture*.

(7) Composed of elements, compounds, or both in variable composition _____

(8) Components do not lose their identity _____

(9) Components separated by chemical changes only _____

(10) Does not resemble elements from which it was formed _____

(11) Composed of two or more elements in fixed proportion by weight _____

(12) Separation may often be made by simple physical or mechanical means _____

8. Fill in the blank space or circle the appropriate response.

The 108 elements, both naturally occurring and artificial, can be classified into three subgroups: metals, nonmetals, and metalloids. Metals, except for mercury, are in the (1) _____ state at room temperature. They have a (2) dull/lustrous appearance and are (3) good/poor conductors of (4) _____ and (5) _____. Metals are also (6) _____, which means they can be drawn into wires, and are (7) _____, which means they can be flattened into sheets. Most metals have (8) _____ melting points and (9) _____ densities.

Nonmetals, in general, have properties (10) similar/opposite to those of metals while elements such as silicon and arsenic, which are called (11) _____, have properties resembling both metals and nonmetals. Metallic elements tend to react with (12) _____ elements to form compounds while non-metallic elements will react with any of the three subgroups. Seven of the non-metallic elements occur naturally as diatomic molecules. Diatomic means that (13) _____ atoms are joined together in chemical bond. Most of the diatomic elements are (14) _____, but one is a liquid and one is a solid at room temperature. Write the formulas for as many of the diatomic elements as you can (15) _____.

9. How many atoms of nitrogen are contained in each of the following formulas?
(1) $NH_4S_2O_8$ _____
(2) NH_4CNS _____
(3) $Na_3Co(NO_2)_6$ _____
(4) $Na_2Fe(CN)_5NO \cdot 2H_2O$ _____
(5) $Fe_2(SO_4)_3(NH_4)_2SO_4 \cdot 24 H_2O$ _____
(6) $K_3Fe(CN)_6$ _____
(7) $NH_4C_2H_3O_2$ _____

10. Name the following binary compounds. Refer to Chapter 8, if necessary.
(1) $HgCl_2$ _____
(2) KCl _____
(3) NaI _____
(4) Al_2S_3 _____
(5) Mg_3N_2 _____
(6) Li_2O _____
(7) SeC_2 _____
(8) CaF_2 _____
(9) AlN _____
(10) AsH_3 _____

11. The following unbalanced chemical equations illustrate possible reactions used to prepare hydrogen gas and oxygen gas. Balance them by placing coefficients in front of the formulas. Don't change the formulas.

   (1) $KClO_3 \rightarrow KCl + O_2$
   (2) $Na + H_2O \rightarrow NaOH + H_2$
   (3) $Zn + HCl \rightarrow ZnCl_2 + H_2$
   (4) $H_2O_2 \rightarrow H_2O + O_2$
   (5) $C + H_2O$ (steam) $\rightarrow CO + H_2$

   Write your answers here.

Challenge Problems

12. 20.0 g of H_2 and 20.0 g of O_2 are reacted to produce H_2O by the following equation:

    $$2H_2 + O_2 \rightarrow 2H_2O$$

    After the reaction, 17.5 g of H_2 are left. What is the percent composition of H_2 and O_2 in H_2O?

    Do your calculations here.

13. Concentrated sulfuric acid (H_2SO_4) has a density of 1.84 g/mL. Dilute sulfuric acid has a density of 1.18 g/mL. For a laboratory experiment you need a sulfuric acid solution with a density of 1.50 g/mL. Using 50 mL of 1.84 g/mL sulfuric acid, how many mL of 1.18 g/mL acid will you need to mix in to obtain the final mixture of 1.50 g/mL?

    Do your calculations here.

RECAP SECTION

Chapter 4 has introduced us to chemical symbols and formulas so that we can now begin speaking the chemist's language. We added important new terms such as *atom, molecule, cation,* and *anion* to our vocabulary, and we know more about the characteristics of pure chemical compounds and mixtures of compounds. Finally, we learned that pure compounds always have the same percentage composition by weight. This is the basis of the chemical Law of Definite Composition, an extremely important summary statement about all chemical compounds.

ANSWERS TO QUESTIONS AND SOLUTIONS TO PROBLEMS

1. (1) elements   (2) compounds   (3) elements   (4) elements
   (5) atom   (6) molecules   (7) molecule   (8) ions
   (9) cation   (10) anion   (11) molecules
   (12) Definite Composition   (13) constant

2. Earth: O, Si, Al, Fe, Ca
   Body: O, C, H, N, Ca, P
   Oxygen and calcium are common to both.

3. Aluminum — Al   Ar — Argon
   Antimony — Sb   Ba — Barium
   Arsenic — As   Bi — Bismuth
   Boron — B   Br — Bromine
   Cadmium — Cd   Cl — Chlorine
   Calcium — Ca   Cr — Chromium
   Carbon — C   Co — Cobalt
   Copper — Cu   F — Fluorine
   Gold — Au   I — Iodine
   Helium — He   Mn — Manganese
   Iron — Fe   Ne — Neon
   Lead — Pb   Ni — Nickel
   Magnesium — Mg   P — Phosphorus
   Mercury — Hg   Si — Silicon
   Nitrogen — N   Na — Sodium
   Potassium — K   S — Sulfur
   Silver — Ag   Ti — Titanium
   Tin — Sn   U — Uranium
   Tungsten — W
   Zinc — Zn

4. (1) No   (2) Yes, molecular   (3) No   (4) Yes, ionic   (5) Yes, molecular
   (6) Yes, molecular   (7) No   (8) No   (9) No   (10) Yes, molecular

5. (1) 2 atoms H, 1 atom S, 4 atoms O
   (2) 1 atom Na, 1 atom O, 1 atom H
   (3) 2 atoms K, 2 atoms Cr, 7 atoms O
   (4) 1 atom Ca, 2 atoms O, 2 atoms H
   (5) 4 atoms P, 10 atoms O
   (6) 3 atoms N, 12 atoms H, 1 atom P, 4 atoms O

(7) 1 atom K, 1 atom Mn, 4 atoms O
(8) 2 atoms Na, 2 atoms H, 2 atoms C, 6 atoms O
(9) 1 atom Cu, 1 atom S, 4 atoms O
(10) 6 atoms Fe, 9 atoms O

6. P_2O_5 and Fe_2O_3

7. (1) Definite Composition   (2) varying
   (3) chemical   (4) physical
   (5) has not   (6) does not
   (7) M   (8) M
   (9) C   (10) C
   (11) C   (12) M

8. (1) solid   (2) lustrous
   (3) good   (4) heat
   (5) electricity   (6) ductile
   (7) malleable   (8) high
   (9) high   (10) opposite
   (11) metalloids   (12) nonmetallic
   (13) two   (14) gases
   (15) $H_2, N_2, O_2, F_2, Cl_2, Br_2, I_2$

9. (1) 1   (2) 2   (3) 6   (4) 6   (5) 2   (6) 6   (7) 1

10. (1) Mercury (II) chloride   (2) Potassium chloride
    (3) Sodium iodide   (4) Aluminum sulfide
    (5) Magnesium nitride   (6) Lithium oxide
    (7) Selenium Carbide   (8) Calcium fluoride
    (9) Aluminum nitride   (10) Arsenic hydride

11. (1) $2 KClO_3 \rightarrow 2 KCl + 3 O_2$ (4) $2 H_2O_2 \rightarrow 2 H_2O + O_2$
    (2) $2 Na + 2 H_2O \rightarrow 2 NaOH + H_2$ (5) $C + H_2O \text{ (steam)} \rightarrow CO + H_2$
    (3) $Zn + 2 HCl \rightarrow ZnCl_2 + H_2$

12. 11% H_2 and 88.9% O_2
    From the data given, the weight of water formed is the amount of O_2 present plus the amount of H_2 reacted.

    $$20.0 \text{ g } O_2 + 2.5 \text{ g } H_2 = 22.5 \text{ g } H_2O$$

    $$\% H_2 = \frac{\text{wt. } H_2}{\text{wt. } H_2O} \times 100 = \frac{2.5 \text{ g}}{22.5 \text{ g}} \times 100 = 11\%$$

    $$\% O_2 = \frac{\text{wt. } O_2}{\text{wt. } H_2O} \times 100 = \frac{20.0 \text{ g}}{22.5 \text{ g}} \times 100 = 88.9\%$$

13. 53 mL of 1.18 g/mL sulfuric acid required. Using the density equals mass divided by volume equation we can solve the problem either by looking at the masses of sulfuric acid involved or look at the densities. Let's use the mass approach.

We can say that the following relationship exists:

(g of concentrated H_2SO_4) + (g of dilute H_2SO_4) = g of mixture and

$$\text{density} = \frac{\text{mass}}{\text{volume}} \text{ or mass} = (\text{density})(\text{volume})$$

Let x = mL of 1.18 g/mL of sulfuric acid.

Therefore g of con. H_2SO_4 = (1.84 g/mL)(50 mL) = 92 g
g of dil. H_2SO_4 = (1.18 g/mL)(x mL) = 1.18 g x

This term contains the unknown, i.e., how much dilute H_2SO_4 is needed.

The density of the mixture is to be 1.50 g/mL so we can set up the following equation.

(1.84 g/mL × 50 mL) + (1.18 g/mL × x mL) = (50 + x)mL × 1.50 g/mL
(final volume) (final density)

Multiply, collect like terms

92 + 1.18(x)mL = 75 + 1.50 (x)mL

17 = 0.32(x)mL

$$x = \frac{17}{0.32} \text{ mL} = 53 \text{ mL}$$

CHAPTER FIVE
Atomic Theory and Structure

SELF-EVALUATION SECTION

1. Fill in the blank space or circle the appropriate response.

   Chapter 5 of the text introduces a number of new terms that will be important in Chapter 6 where we study a system of arranging the elements in a periodic fashion. As we begin to characterize various elements, one definition should automatically come to mind. The number of protons in an atom of any element is called the (1) _____ _____. This number is also the same as the number of (2) _____ in a neutral atom. This is logical since in an electrically neutral atom the number of positive charges has to equal the number of (3) _____ charges.

   The two subatomic particles which have a mass of 1 amu are the (4) _____ and (5) _____. The negatively charged particle, called an (6) _____, has an assigned mass of (7) _____. Thus, essentially all of the mass of an atom is located in the (8) _____ of an atom. In describing the relative masses of the atoms of various elements, we use the term (9) _____ _____, which represents the mass of the number of protons and neutrons in an atom. However, as we look at a list of relative atomic weights for various elements, we find that most values are not whole numbers. This may seem odd since the system of atomic weights is based on a particular isotope of the element, (10) _____, having an exact atomic weight of 12.00. We must remember that atoms of an element can have different numbers of neutrons in the nucleus and still be the same element. Thus, a carbon atom can have six neutrons or eight neutrons in the nucleus. The atom is still carbon since the atomic number remains 6, but the relative mass of the two atoms is different. And, depending on the relative abundance of the various atoms with differing numbers of neutrons, the average atomic weight of a sample of the element will not be a whole number.

The atomic weight of the element is a weighted average of all the atoms present in the sample.

Atoms of a particular element which have the same atomic (11) _____ but differ in atomic mass are called (12) _____ of that element. Some radioactive isotopes are extremely important today. Some such as $^{131}_{53}I$, $^{60}_{27}Co$, or $^{32}_{15}P$ are used in medical applications. Others, such as $^{90}_{38}Sr$, are environmental contaminants and may pose a long-term problem for various organisms. The simplest element, hydrogen, has three isotopes. They are named (13) _____, (14) _____, and (15) _____.
The symbols for the elements used in the preceding sentences have two numbers in addition to the chemical symbol. The upper number, which is the larger, is the (16) _____ _____. The lower number is the (17) _____ _____.

2. Indicate by a yes or no response which of the following statements are generally accepted today.

   (1) Atoms of the same element may vary in mass but are the same size. _____

   (2) Compounds are formed by the union or joining of two or more atoms of different elements. _____

   (3) Atoms of different elements have different masses and sizes. _____

   (4) Atoms of two elements may combine in different ratios to form more than one compound. _____

   (5) Elements are composed of small, indivisible units called electrons. _____

   (6) In forming compounds, atoms can combine in fractional units. _____

   (7) The fundamental building blocks of elements are small particles called atoms. _____

   (8) Atoms of the same element are alike in mass and size. _____

   (9) In forming compounds, atoms combine in small whole-number ratios such as one to one, one to two, and so forth. _____

   (10) Atoms of two elements can combine in only one fixed ratio to form only one compound. _____

3. Description of properties of subatomic particles. Fill in the spaces of the table with the correct response from the list of possible responses below.

Particle	Symbol	Relative Mass (amu)	Charge	Location in Atom
Electron				
Proton				
Neutron				

Possible responses

Symbol	Relative Mass (amu)	Charge	Location in Atom
ne	+2	−1	nucleus
p	+1	−2	electron orbital
e$^+$	−1	+1	
e$^-$	−1/2	−2	
n	0	0	
pr	1/1837	−1/2	

4. Fill in the blank space or circle the appropriate response.

The three principal subatomic particles which are of interest to chemists are the (1) _____, (2) _____, and (3) _____. A series of experiments after the turn of the century showed that protons and (4) _____ were found in the (5) _____ of an atom while the (6) _____ were in regions outside of the nucleus. Rutherford demonstrated that practically all of the mass of an atom is to be found in the (7) _____ and that atoms are mostly empty space. The experiments of Rutherford and others left unanswered questions about the nature of electrons in atoms. Niels Bohr, a Danish physicist, suggested that the electron for the simplest atom, hydrogen, didn't wander randomly about the nucleus, but moved only in certain well-defined orbits. You may have seen an atomic symbol used by a power company or other agency illustrating this model, which is analogous to orbits of planets around the sun. Bohr's theory states that there are various well-defined orbits available to an electron, some close to the nucleus and some farther away. When an electron absorbs energy, it jumps to a (8) higher/lower energy level, and as it loses energy, it moves to a (9) higher/lower energy level; the excess energy is given off as light energy. Bohr's theory explains very nicely the atomic spectra of hydrogen, but it does not explain spectra of more complicated multielectron atoms. The quantum mechanic theory has replaced Bohr's theory today as the best explanation for electron behavior. The significant difference in the two theories for us is that quantum mechanics states that it is impossible to know exactly the position of an electron at any instant in time. What we can say and visually represent is that electrons occupy orbitals around the nucleus which are the regions of space where the electrons most probably are. We often use the term *electron cloud* to describe the region of space defined by the various energy levels. The word *cloud* is appropriate because it pictures the somewhat nebulous behavior of an electron. Quantum or wave mechanic equations provide us with a tool to describe the probable location of electrons in an atom.

5. Write the electronic configuration for the elements listed, using the basic rules and energy level order regarding the state of electrons in atoms listed as a, b, and c.
   a. In the ground state (lowest energy state) of an atom, electrons tend to occupy orbitals of the lowest possible energy.
   b. Each orbital may contain a maximum of two electrons (with opposite spins).
   c. The energy level order is

   $1s, 2s, 2p, 3s, 3p, 4s, 3d$

   The maximum number of electrons found in any main energy level can be determined by the formula $2n^2$ where n is the (1) _____

   _____. Thus, for $n = 1$, the total number of electrons will be (2) _____, which will fill the suborbitals (3) s, p, d, f. For $n = 2$, the total number of electrons will be (4) _____, which will fill the suborbitals (5) s, p, d, f. Look at the following symbol and describe its meaning:

   $4s^2$

   The number 4 represents (6) _____, the small letter s represents (7) _____, and the superscript 2 represents (8) _____.

   Now we can write our electronic configurations. For example, lithium has atomic number 3. Therefore, the correct way of writing the electronic configuration would be Li $1s^2 2s^1$.

   Try the configurations for   (9) Nitrogen   Atomic number 7
                                             (10) Argon   Atomic number 18
                                             (11) Sodium   Atomic number 11
                                             (12) Iron   Atomic number 26
                                             (13) Calcium   Atomic number 20

   Write your answers here.

Determine the atomic number for the element represented by the following electronic configurations.

(14) $1s^2 2s^2 2p^6 3s^2 3p^2$ _____

(15) $1s^2 2s^2 2p^4$ _____

(16) $1s^2 2s^2 2p^6 3s^2 3p^6 4s^2 3d^7$ _____

(17) $1s^2 2s^2 2p^6 3s^2 3p^6 4s^2 3d^1$ _____

(18) $1s^2 2s^2 2p^6 3s^2 3p^6 4s^2$ _____

6. Atomic weights, atomic numbers, and number of neutrons.
   (1) Determine the number of neutrons in an atom of each of the following elements.

   $^{23}_{11}Na$ _____ $^{65}_{30}Zn$ _____

   $^{115}_{48}Cd$ _____ $^{40}_{18}Ar$ _____

   $^{35}_{17}Cl$ _____ $^{40}_{20}Ca$ _____

   (2) Determine the approximate *atomic weight* for an atom of each of the following isotopes.

	Atomic Number	Number of Neutrons	Atomic Weight
Li	3	4	_____
K	19	20	_____
Hg	80	121	_____
Fe	26	30	_____
H	1	0	_____
P	15	16	_____

7. Using Table 5.5 in your textbook, construct electron-dot diagrams for the following representative atoms.

   (1) Potassium
   (2) Arsenic
   (3) Iodine
   (4) Phosphorus
   (5) Bromine
   (6) Zinc
   (7) Copper
   (8) Aluminum
   (9) Carbon
   (10) Sulfur

For example, Table 5.5 gives the electronic structure of magnesium as [Ne] $3s^2$, which means that Mg has two outer orbital electrons to show in the dot diagram.

Mg:

Write your answers here.

8. Which of the following statements are true about the neutral atoms $^{12}_{6}C$ and $^{14}_{6}C$?

   (1) $^{12}_{6}C$ and $^{14}_{6}C$ are isotopes.
   (2) $^{12}_{6}C$ has two less electrons than $^{14}_{6}C$.
   (3) $^{14}_{6}C$ has two more protons than $^{12}_{6}C$.
   (4) $^{12}_{6}C$ has two less neutrons than $^{14}_{6}C$.

Challenge Problems

9. After studying figure 5.7 for a few minutes, try your hand at constructing the orbital electron structure ($1s$, $2s$, etc.) for (1) $_{21}Sc$ (2) $_{29}Cu$ (3) $_{40}Zr$.

   Write your answers here.

RECAP SECTION

The discussion in Chapter 5 centers around the topic of atomic structure and what we know about the composition of an atom. Considering that most of the experiments in atomic structure have been done since 1900, we have gained a great deal of knowledge. Now we are able to explain atomic spectra and the existence of isotopes. The special stability of the noble gases can be attributed to their atoms having 8 electrons in the outer energy level (i.e., the s and p orbitals of the outer energy level are filled and contain 8 electrons, s^2p^6).

ANSWERS TO QUESTIONS AND SOLUTIONS TO PROBLEMS

1. (1) atomic number (2) electrons (3) negative
   (4) proton (5) neutron (6) electron
   (7) 1/1837 amu (8) nucleus (9) atomic weight
   (10) carbon (11) number (12) isotopes
   (13) protium (14) deuterium (15) tritium
   (16) mass number (17) atomic number

2. (1) yes (2) yes (3) yes (4) yes (5) no
   (6) no (7) yes (8) no (9) yes (10) no

3.

Particle	Symbol	Relative Mass (amu)	Charge	Location in Atom
Electron	e	1/1837	−1	electron orbital
Proton	p	1	+1	nucleus
Neutron	n	1	0	nucleus

4. (1) protons (2) neutrons (3) electrons (4) neutrons
   (5) nucleus (6) electrons (7) nucleus (8) higher
   (9) lower

5. (1) principal energy level
   (2) 2 (3) s (4) 8 (5) s, p (6) principal energy level
   (7) Type of sublevel (8) Number of electrons in that sublevel
   (9) $1s^2 2s^2 2p^3$ (10) $1s^2 2s^2 2p^6 3s^2 3p^6$ (11) $1s^2 2s^2 2p^6 3s^1$
   (12) $1s^2 2s^2 2p^6 3s^2 3p^6 4s^2 3d^6$ (13) $1s^2 2s^2 2p^6 3s^2 3p^6 4s^2$
   (14) 14 (15) 8 (16) 27 (17) 21 (18) 20

6. (1) Na = 12   Zn = 35
       Cd = 67   Ar = 22
       Cl = 18   Ca = 20

   (2) Li = 7   Fe = 56
       K = 39   H = 1
       Hg = 201   P = 31

7. (1) K·   (2) ·A̤s·   (3) ·Ï: (4) ·P̤·

   (5) ·B̈r: (6) Zn: (7) Cu·   (8) Al:

   (9) ·C̣:   (10) ·S̈:

8. (1) T   (2) F   (3) F   (4) T

9. (1) $_{21}$Sc $1s^2$ $2s^2$ $2p^6$ $3s^2$ $3p^6$ $4s^2$ $3d^1$
   (2) $_{29}$Cu $1s^2$ $2s^2$ $2p^6$ $3s^2$ $3p^6$ $4s^1$ $3d^{10}$
   (3) $_{40}$Zr $1s^2$ $2s^2$ $2p^6$ $3s^2$ $3p^6$ $4s^2$ $3d^{10}$ $4p^6$ $5s^2$ $4d^2$

CHAPTER SIX
The Periodic Arrangement of the Elements

SELF-EVALUATION SECTION

1. Fill in the blank space or circle the appropriate response.

    The periodic law states that the properties of elements are periodic functions of their atomic (1) _____. The word *periodicity* suggests that elemental properties in the periodic table occur (2) <u>only once/repeatedly</u>. If the known elements are arranged in a table according to increasing atomic number rather than increasing atomic (3) _____, we can begin to talk about various groupings of elements. A horizontal group of elements in the periodic table is called a (4) _____. A vertical group of elements is called a (5) _____. Each period of elements is numbered with an integer that indicates the outermost (6) _____ being filled progressively with electrons. The members of a group or (7) _____ of elements are related to each other in that the outermost energy level of each contains (8) <u>the same/a different</u> number of electrons. Because of the electronic configurations, groups and families of elements will have (9) <u>similar/different</u> chemical properties. Beginning with the 4th period, (10) <u>terminal/transition</u> elements occur in the (11) <u>middle/end</u> of the period. These elements are all (12) <u>nonmetals/metals</u> and form a variety of compounds. The distinguishing characteristic of the transition elements is that an (13) <u>inner/outer</u> shell of electrons is being filled as the atomic number increases. The distinguishing electrons for the transition elements are (14) <u>s & p/d & f</u>.

    Focusing our attention on the various groups or families of elements, we can make some general statements concerning group characteristics. For Groups IA-VIIA, IB, and IIB, the group number is the same as the number of (15) _____ in the outer energy level. The other groups of elements, the transition metals, are filling (16) <u>inner/outer</u> electronic orbitals and cannot be classified as representative elements. In addition, the noble gases (with the exception of helium) have (17) _____ electrons in their outer energy level.

The elements on the left side of the periodic table are (18) metals/nonmetals and share the general characteristics of all metals. The elements at the top of the groups on the right side of the periodic table are (19) metals/nonmetals. However, as one proceeds down these groups toward the elements with larger nuclei, the metallic characteristics of the elements (20) increase/decrease. Thus, Group VA begins with nitrogen, a nonmetal, and ends with bismuth, a metal. In general, as one goes across a period the atomic radii (21) decrease/increase as the atomic number increases.

Since all elements in a group have (22) different/the same outer orbital electron structure, one can expect that elements in a group will have (23) similar/varied chemical properties. For example, if magnesium (Mg) forms a compound with oxygen with the formula MgO, then a compound formed from barium (Ba), which is in the same group as magnesium, and oxygen, would have the formula (24) _____. In another example, aluminum oxide has a formula of Al_2O_3. Sulfur, which is in the same group as oxygen, forms a compound with aluminum named aluminum sulfide, which has the formula (25) _____.

2. Using the periodic table in Chapter 6 of the textbook, find the following items of information about each element.

	Symbol	Atomic Number	Atomic Weight	No. of Outer Orbital Electrons
(1) Phosphorus	_____	_____	_____	_____
(2) Fluorine	_____	_____	_____	_____
(3) Mercury	_____	_____	_____	_____
(4) Cesium	_____	_____	_____	_____

3. Many scientists were involved in the development of the periodic table. Match the names with the descriptive phrases.

   a. Newlands   d. Moseley
   b. Döbereiner   e. Mendeleev
   c. Meyer

   (1) German scientist whose periodic arrangement was
       published closely behind the more famous one   _____

(2) Determined that the atomic number of an element placed the element in periodic order _____

(3) Triads _____

(4) Law of Octaves _____

(5) Russian scientist whose table had blank spots for yet-undiscovered elements _____

4. Identify each of the following representative elements as being an alkali metal, alkaline earth, halogen, or noble gas. Give the Lewis dot symbol for each representative element also using the periodic table.

   (1) Calcium
   (2) Argon
   (3) Iodine
   (4) Potassium
   (5) Barium
   (6) Chlorine
   (7) Cesium
   (8) Helium
   (9) Magnesium
   (10) Lithium
   (11) Bromine
   (12) Xenon

Challenge Problems

5. One of the earlier attempts to classify the elements was by the German chemist J.W. Döbereiner using the table of atomic weights published by Berzelius in 1826. Döbereiner grouped certain elements into groups of threes or triads. The middle element of the triad had an atomic weight that was the mean of the atomic weights of the lighter and the heavier element. For example, one triad consisted of the elements Li, Na, K. Atomic weights are Li = 6.9, Na = 23, K = 39.1. The mean $\frac{6.9 + 39.1}{2}$ is 23 which is the accepted atomic weight of the middle element Na. From your periodic table you can see that all three of these elements are Group I elements.

   Now, using your table of atomic weights in the text and Döbereiner's calculation method, determine the atomic weight and identity of each of the following members of a triad.

   (1) Beryllium, Unknown "A", Calcium

   (2) Chlorine, Unknown "B", Iodine

   (3) Sulfur, Unknown "C", Tellurium

   (4) Potassium, Unknown "D", Cesium

   Write your answers in the spaces provided.

	Identification	Calculated mean atomic weight	Accepted at. wt.
Unknown "A"			
Unknown "B"			
Unknown "C"			
Unknown "D"			

6. Mendeleev, in formulating his periodic arrangement of the elements, was able to predict in 1871 the properties of two as yet undiscovered elements: eka-silicon and eka-aluminum. Using reference sources such as the Handbook of Chemistry and Physics (Chemical Rubber Co.), and a periodic table, fill in the following table and answer the questions.

	eka-Aluminum	eka-Silicon
Atomic weight predicted	68	72
actual		
Specific gravity predicted	5.9	5.5
actual		
Specific heat predicted	0.091 cal/g	0.073 cal/g
actual		
Identification of element		

The predicted oxidation states for eka-aluminum were +2 and +3 and for eka-silicon +2 and +4. With this in mind, write the following formulas using the negatively charged ions listed along the top of the table and the correct accepted modern symbols for the two positive elements.

	Cl^-	O^{2-}	S^{2-}
Gallium (+3)			
Germanium (+2)			
Gallium (+2)			
Germanium (+4)			

RECAP SECTION

The material in Chapter 6 has introduced us to a most important practical tool, the periodic table. The information contained in the complete long form of the periodic table is used by scientists everywhere during their day-to-day work. As you continue on in science, you will find yourself referring to the table to calculate molecular weights of chemicals, to look up physical properties, and to determine oxidation states and theoretically possible chemical formulas.

ANSWERS TO QUESTIONS

1. (1) number (2) repeatedly (3) weight (4) period
   (5) group or family (6) energy level (7) family (8) the same
   (9) similar (10) transition (11) middle (12) metals
   (13) inner (14) d & f (15) electrons (16) inner
   (17) eight (18) metals (19) nonmetals (20) increase
   (21) decreases (22) the same (23) similar (24) BaO
   (25) Al_2S_3

2. (1) P   15   31.0   5
   (2) F   9   19.0   7
   (3) Hg 80   200.6 2
   (4) Cs 55   132.9 1

3. (1) c   (2) d   (3) b   (4) a   (5) e

4. (1) alkaline earth, $\cdot\ddot{Ca}\colon$ (7) alkali metal, $Cs\cdot$
   (2) noble gas, $\colon\ddot{Ar}\colon$ (8) noble gas, $He\colon$
   (3) halogen, $\cdot\ddot{\underset{..}{I}}\colon$ (9) alkaline earth, $\cdot\ddot{Mg}\colon$
   (4) alkali metal, $K\cdot$ (10) alkali metal, $Li\cdot$
   (5) alkaline earth, $\cdot\ddot{Ba}\colon$ (11) halogen, $\cdot\ddot{\underset{..}{Br}}\colon$
   (6) halogen, $\cdot\ddot{\underset{..}{Cl}}\colon$ (12) noble gas $\colon\ddot{\underset{..}{Xe}}\colon$

5.

	Identification	Calculated mean atomic weight	Accepted at. wt.
Unknown "A"	Mg	24.6	24.3
Unknown "B"	Br	81.2	79.9
Unknown "C"	Se	79.8	79.0
Unknown "D"	Rb	86	85.5

6.

	eka-Aluminum	eka-Silicon
Atomic weight actual	69.7	72.6
Specific gravity	5.9	5.3
Specific heat	0.088 cal/g	0.077 cal/g
Identification	Gallium	Germanium

Formulas

$GaCl_3$	Ga_2O_3	Ga_2S_3
$GeCl_2$	GeO	GeS
$GaCl_2$	GaO	GaS
$GeCl_4$	GeO_2	GeS_2

CHAPTER SEVEN
Chemical Bonds—The Formation of Compounds from Atoms

SELF-EVALUATION SECTION

1. Fill in the blank space or circle the appropriate response.

When a neutral atom loses an orbital electron, it becomes a (1) <u>positively/negatively</u> charged ion. The energy required to remove a mole of electrons from a mole of atoms is called the (2) _____ energy and is a relatively (3) <u>low/high</u> value for Group I and Group II metals and (4) <u>low/high</u> for nonmetallic elements. The ionization energy for the noble gas elements is especially (5) <u>high/low</u>, indicating that eight electrons in the outer shell of an atom is a very stable structure. If you list the ionization energies of the elements from a group in the periodic table, the element from the top of the group has a (6) <u>higher/lower</u> ionization energy than the element at the bottom of the group. Two factors account for this experimental observation. As you go down a group, the electron being removed is (7) <u>farther from/closer to</u> the nucleus, and the increasing number of filled electron orbitals shields the outermost electrons from the positive nucleus. The outermost electrons, which are also called (8) _____ electrons, are shown in the electron-dot structures. We will use electron-dot structures in the next section to illustrate the concept of forming chemical bonds.

As mentioned above, when a neutral atom (9) <u>loses/gains</u> an electron it becomes positively charged. A positively charged ion is also called a (10) _____. For example, a potassium atom has 19 protons and 19 electrons. A potassium ion has a charge of +1, which means the ion has (11) _____ electrons instead of 19, as in the neutral atom. Conversely, a negatively charged ion, which is called an (12) _____, is formed when a neutral atom (13) <u>loses/gains</u> electrons. A chlorine atom has 17 protons and 17 electrons and becomes an anion by (14) <u>gaining/losing</u> an electron. A chloride ion has a stable outer shell of (15) _____ electrons.

A sodium atom in close proximity to a chlorine atom can also reach a stable outer shell of eight electrons by losing one electron. Thus, both sodium and chlorine reach a stable electron structure by the process of electron transfer. The metallic elements attain a stable structure by (16) gaining/losing electrons; the nonmetallic elements attain a stable structure by (17) gaining/losing electrons.

2. In this section, we will use electron-dot structures to illustrate how chemical bonds are formed. Using the techniques in Sections 7.4 and 7.5, we will work out some electron-transfer problems, then some dealing with electron sharing.

   (1) Using electron-dot structures, show how the compound zinc iodide (ZnI_2) is formed. Zinc is a Group IIB element and iodine is a Group VIIA element.

   (2) Show how the compound potassium oxide (K_2O) is formed. Potassium is a Group IA element, and oxygen is a Group VIA element.

   (3) Let's try some problems that involve electron sharing between atoms to obtain a stable configuration of eight electrons. Show how the compound phosphorus trichloride (PCl_3) is formed. Phosphorus is a Group VA element, and chlorine is in Group VIIA.

   (4) Show how hydrogen sulfide (H_2S) is formed. Hydrogen is a Group IA element, and sulfur is a Group VIA element.

(5) Show how the compound carbon dioxide (CO_2) is formed. Carbon is a Group IVA element, and oxygen is a Group VIA element. Be careful that each atom has eight electrons through sharing. Consider the possibility of using double bonds if you don't have enough electrons for all single bonds.

(6) Draw the electron-dot structure for SO_3. Sulfur and oxygen are both Group VIA elements.

3. Write out the Lewis structure for HNO_2, nitrous acid. Point out the covalent and coordinate bonds. You may use a periodic table and follow problems in the text as a guide.

4. Draw Lewis structures for the following:
   (1) NH_4Cl (2) PO_4^{3-} (3) H_2O
   (4) MnO_4^- (5) CO_3^{2-} (6) ClO^-

5. If we examine many different compounds for the kinds of chemical bonds that hold them together, we will generally find two types. The two types of chemical bonds are called the (1) _____ and the (2) _____ bond.

When a transfer of electrons takes place from one atom to another, an (3) _____ bond is formed. A cation and an anion will form an (4) _____ bond since oppositely charged particles (5) attract/repel each other. Metallic elements tend to form ionic bonds when combining with the nonmetals, as we saw in the previous section.

The predominant type of chemical bond is the (6) _____ bond. This type of bond occurs in the hydrogen molecule and develops as a result of each hydrogen atom contributing (7) <u>one/two</u> electron(s) to form the bond. The 1s electron orbitals of the hydrogen atoms overlap and pair-up to form a stable hydrogen molecule. There is a strong tendency for the hydrogen molecule to form from two individual atoms since in the molecule each (8) _____ charged electron is attracted to two (9) _____ charged nuclei. The covalent bond is usually indicated by a dash mark (—). A single dash means (10) _____ pair of electrons and a double dash means (11) _____ pairs of electrons.

6. Given the following table of electronegativity values, indicate which of the listed binary compounds has polar covalent bonds. Also calculate the difference in electronegativity values for one of the covalent bonds in each molecule.

Electronegativity Values

H	2.1	Br	2.8	Se	2.4
B	2.0	Cl	3.0	Te	2.1
P	2.1	F	4.0	N	3.0
O	3.5	S	2.5	Na	0.9

For example, the compound HF has one covalent bond. Is the bond between H and F polar? Yes, there is a significant difference in electronegativity values of $4.0 - 2.1 = 1.9$. Examine the remaining compounds in the same manner.

Compound	Polar Covalent Bond (yes or no)	Electronegativity Value Difference
(1) H_2O	_____	_____
(2) PH_3	_____	_____
(3) BrCl	_____	_____
(4) H_2S	_____	_____
(5) OF_2	_____	_____
(6) NH_3	_____	_____
(7) H_2Se	_____	_____
(8) H_2Te	_____	_____
(9) Na_3P	_____	_____
(10) NCl_3	_____	_____
(11) Br_2	_____	_____

7. In the following, which of each pair will be larger?
    (1) Cl^- and Cl^0
    (2) Al^0 and Al^{3+}
    (3) Na^+ and Na^0
    (4) Fe^{2+} and Fe^{3+}

8. Draw Lewis structures for the following:
    (1) H_2CO
    (2) $S_2O_3^{2-}$

9. Predict the type of bond that would be formed between the following pairs of atoms. Use Table 7.3 in the text.
    (1) Li & I
    (2) C & H
    (3) Al & N
    (4) Cs & P
    (5) Se & S
    (6) Ba & C

RECAP SECTION

Chapter 7 presents you with a great deal of useful information about the formation of chemical compounds from individual atoms. You now know how to describe chemical bonding in terms of electron transfer or electron sharing. And by using the concept of electronegativity, you can be more precise about the kind of electron sharing that takes place in covalent bonding. The concept of electronegativity is a most valuable tool in chemistry. The general theme of the unit has been to examine the formation of compounds and to describe the forces of attraction found in compounds.

ANSWERS TO QUESTIONS AND SOLUTIONS TO PROBLEMS

1. (1) positively   (2) ionization   (3) low   (4) high
   (5) high   (6) higher   (7) farther from   (8) valence
   (9) loses   (10) cation   (11) 18   (12) anion
   (13) gains   (14) gaining   (15) eight   (16) losing
   (17) gaining

2. (1) Zinc is a Group IIB element, which means that it has two valence electrons in its outer shell. Iodine is a Group VIIA element, which means that it has seven valence electrons in its outer shell. Zinc loses one electron to each of the iodine atoms and becomes a +2 charged cation. Each iodide ion thus has a −1 charge.

$$Zn + \overset{..}{\underset{..}{I}}: \; \overset{..}{\underset{..}{I}}: \longrightarrow Zn^{2+} \; \begin{matrix}:\overset{..}{\underset{..}{I}}:^{-} \\ :\overset{..}{\underset{..}{I}}:^{-}\end{matrix}$$

(2) $K \cdot \; + \; \overset{..}{\underset{..}{O}}: \; + \; \cdot K \longrightarrow K^{+} \; :\overset{..}{\underset{..}{O}}:^{2-} \; K^{+}$

(3) $:\overset{..}{\underset{..}{Cl}}\cdot \longrightarrow \cdot \overset{.}{P} \cdot \longleftarrow \cdot \overset{..}{\underset{..}{Cl}}: \longrightarrow :\overset{..}{\underset{..}{Cl}}: \overset{:\overset{..}{\underset{..}{Cl}}:}{P} :\overset{..}{\underset{..}{Cl}}:$
   with $:\overset{..}{\underset{..}{Cl}}:$ below P

(4) $H \cdot \longrightarrow \cdot \overset{.}{\underset{..}{S}}: \longleftarrow H \longrightarrow H : \overset{..}{\underset{..}{S}}:$
   with H below S

(5) In order to write an electron-dot structure for the molecule CO_2, which will have eight electrons around each atom, we must use double bonds between C and each oxygen atom. Each double bond consists of four electrons—two from the C atom and two from each oxygen atom.

$$:\overset{..}{\underset{..}{O}} \longrightarrow \cdot \overset{.}{\underset{.}{C}}: \; \overset{..}{\underset{..}{O}}: \longrightarrow :\overset{..}{\underset{..}{O}}::C::\overset{..}{\underset{..}{O}}:$$

The problem involves the formation of double bonds. Since one pair of electrons is one bond, there are two bonds between the two oxygen atoms and the carbon atom. Only in this manner will each atom have eight electrons in its outer shell.

(6) $S + :\overset{..}{\underset{..}{O}} \; \overset{..}{\underset{..}{O}}: \; \overset{..}{\underset{..}{O}}: \longrightarrow \begin{matrix}:\overset{..}{\underset{..}{O}}: \\ S \\ :\overset{..}{\underset{..}{O}}: \; :\overset{..}{\underset{..}{O}}:\end{matrix}$

The dot structure for SO_3 involves one double bond and two single bonds in order to place eight electrons around each atom.

3. H:Ö:N::Ö: or   H—O—N=O
   all covalent bonds   a "dash" equals a pair of electrons

Working out a Lewis structure for a somewhat complicated compound such as HNO_2 is a little bit like working a puzzle. We have four pieces to fit together so that each piece has a complete "set" of electrons—either two (as for H) or eight (as for O and N). Let's start by listing the number of electrons available.

N has five electrons, each O has six electrons, and H has one electron.

Take nitrogen first. We need three electrons to fill out the eight. It's common for oxygen to share one or two electrons with another atom.

If one O shares two electrons with nitrogen, then the other oxygen can share one with hydrogen and another with the nitrogen. Using symbols, we have

H:Ö::N:Ö:

We should simplify this to show only two electrons on each of the four sides of the oxygen atoms and the nitrogen atom.

H:Ö:N:Ö:

With this situation the H, one O, and the N are happy with their number of electrons but the second O has only six. We have another possibility, which is to move one of the nonbonded pairs of electrons from N to go between the N and the O.

H:Ö:N::Ö:

Now all four atoms have filled orbitals and the bonds are covalent bonds. Another possibility you might have tried would be to bond the H to the nitrogen.

     H
:Ö::N::Ö:

This structure satisfies the requirements for the number of electrons around each atom but involves two double bonds, one each between the nitrogen and each oxygen. However, other evidence suggests that this is not the correct structure. Also, from other evidence we know that HNO_2 is an acid, which suggests the H is bonded to an oxygen atom.

4. (1)
              H   ⊕
        H:N:H   :Cl:⊖
              H

(2)
              :Ö: 3−
         :Ö:P:Ö:
              :Ö:

57

(3) 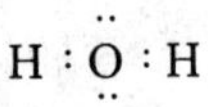 H : Ö : H

(4) 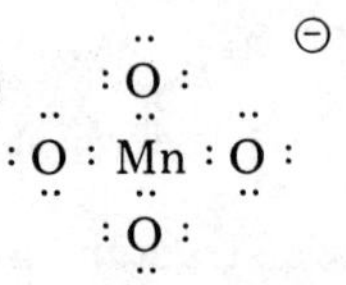
:Ö:
:Ö: Mn : Ö: ⊖
:Ö:

(5) 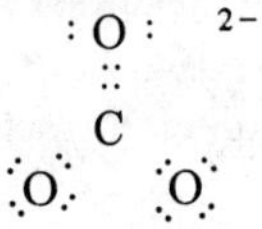 :Ö: 2−
   C
:Ö: :Ö:

There are 4e⁻ from the carbon, 6e⁻ each from the oxygen and 2 additional (negative charge on the ion is −2). So we have to distribute 24e⁻ between the central carbon and the 3 oxygen atoms. This means that one of the bonds needs to be shown as a double bond.

(6) 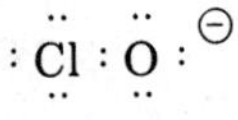 :C̈l : Ö : ⊖

5. (1) ionic   (2) covalent   (3) ionic
    (4) ionic   (5) attract   (6) covalent
    (7) one   (8) negatively (9) positively
    (10) one   (11) two

6. Compound	Polar Covalent Bond (yes or no)	Electronegativity Value Difference
(1) H_2O	yes	1.4
(2) PH_3	no	0
(3) BrCl	yes	0.2
(4) H_2S	yes	0.4
(5) OF_2	yes	0.5
(6) NH_3	yes	0.9
(7) H_2Se	yes	0.3
(8) H_2Te	no	0
(9) Na_3P	yes	1.2
(10) NCl_3	no	0
(11) Br_2	no	0

7. (1) Cl⁻ will be larger.   (3) Na⁰ will be larger.
   (2) Al⁰ will be larger.   (4) Fe²⁺ will be larger.

8.
   (1) H:C::O with H on top of C   (2) $[:\ddot{O}:\ddot{S}:\ddot{O}:\ddot{S}:\ddot{O}:]^{2-}$

9. (1) polar covalent (2) polar covalent (3) polar covalent (4) polar covalent
   (5) non polar covalent (6) ionic

WORD SEARCH 1

In the given matrix of letters, find the terms that match the following definitions. The terms may be horizontal, vertical, or on the diagonal. They may also be written forward or backward. Answers are found on page 235.

1. A subatomic particle with a charge of +1.
2. The mass of an object divided by its volume.
3. The basic building block of matter that cannot be broken down into simpler substances by ordinary chemical changes.
4. An electrically charged atom or group of atoms.
5. The central part of an atom.
6. An element that is ductile and malleable.
7. The abbreviation for the name of an element.
8. The chemical law concerning the occurrence of the chemical properties of elements.
9. State of matter that is least compact.
10. Atoms of an element having same atomic number but different atomic masses.
11. Instrument used to measure the specific gravity of a liquid.
12. A subatomic particle with a charge of −1.
13. A negatively charged ion.
14. A small, uncharged individual unit of a compound.
15. Matter having uniform properties throughout.
16. The smallest particle of an element that can enter into a chemical reaction.
17. The relative attraction that an atom has for the electrons in a covalent bond.
18. A molecule with a separation of charge.
19. A subatomic particle that is electrically neutral.
20. The energy required to remove an electron from an atom.
21. A substance composed of two or more elements combined in a definite proportion by weight.
22. A solid without definite crystalline form.
23. A cloud-like region around the nucleus where electrons are located.
24. The metallic elements characterized by increasing numbers of d and f electrons in an inner shell.
25. Metric unit of length.

H	C	O	M	P	O	U	N	D	D	I	S	O	T	O	P	E	S
O	X	C	S	Z	K	L	O	Y	P	M	N	Q	R	B	L	C	T
M	P	R	U	D	K	Q	T	V	F	V	G	M	W	E	Z	T	Y
O	I	Z	E	O	R	V	O	M	T	S	E	O	C	U	A	L	G
G	U	A	L	T	E	I	R	O	R	E	S	T	T	L	A	C	I
E	I	F	C	S	U	P	P	W	L	L	R	A	B	A	I	T	U
N	G	O	U	Q	E	G	Y	O	H	O	J	Y	G	R	N	O	P
E	E	E	N	E	U	T	R	O	N	P	A	M	V	E	E	S	E
O	L	B	U	I	S	W	H	E	J	I	M	N	M	T	K	O	R
U	S	O	V	T	Z	U	G	J	F	D	V	E	C	L	D	E	I
S	I	E	I	R	K	A	O	A	B	K	L	T	U	M	L	E	O
K	N	G	S	D	T	M	T	H	V	E	P	R	S	D	D	A	D
L	L	A	T	I	B	R	O	I	P	A	D	O	I	M	E	R	I
O	E	W	V	S	L	H	F	L	O	R	G	E	U	Q	N	V	C
B	O	I	O	N	N	A	P	A	E	N	O	L	T	W	S	Z	L
M	T	V	M	E	O	S	I	D	N	C	E	M	N	O	I	N	A
Y	L	S	A	T	R	T	Z	M	A	B	U	N	A	L	T	U	W
S	P	J	D	A	T	C	I	O	T	M	J	L	E	K	Y	V	Z
A	E	O	R	B	C	M	R	P	E	B	U	M	E	R	E	I	D
B	H	T	L	C	E	A	H	T	M	X	Y	U	T	T	G	G	C
G	D	I	U	F	L	Q	A	O	F	E	S	N	X	G	A	Y	V
S	T	N	E	M	E	L	E	N	O	I	T	I	S	N	A	R	T
A	M	G	W	U	P	Q	N	S	W	D	L	E	R	V	K	C	B
H	I	V	R	E	T	E	M	O	R	D	Y	H	R	W	E	J	F

CHAPTER EIGHT
Nomenclature of Inorganic Compounds

SELF-EVALUATION SECTION

1. Write the correct names for the chemical formulas as listed. For additional assistance in naming compounds, refer to Tables 8.2, 8.3, and 8.4.

   (1) KF K_2O KNO_3 K_3PO_4

   _____

   (2) MgF_2 MgO $Mg(NO_3)_2$ $Mg_3(PO_4)_2$

   _____

   (3) FeF_3 Fe_2O_3 $Fe(NO_3)_3$ $FePO_4$

   _____

   (4) HF H_2O HNO_3 H_3PO_4

   _____

   (5) SnF_4 SnO_2 $Sn(NO_3)_4$ $Sn_3(PO_4)_4$

   _____

2. Write the names of the following compounds, which are composed of nonmetallic elements covalently bonded together.

   (1) CO CO_2 PCl_5 P_2O_5

   _____

   (2) CCl_4 NO S_2Cl_2 N_2O_3

   _____

3. There are some common exceptions to the general rules that state that only binary compounds end in *ide*. Name the compounds listed.

   (1) NH_4Br _____ $(NH_4)_2S$ _____

   (2) KOH _____ $Mg(OH)_2$ _____

   (3) NaCN _____ HCN _____

4. Write the correct names for these common compounds.

   (1) Common acids

   HCl H_2SO_4 $HC_2H_3O_2$ H_2CO_3

   _____   _____   _____   _____

   (2) Common salts

   $MgSO_4 \cdot 7 H_2O$ K_2CO_3 $NaNO_3$ $Ca_3(PO_4)_2$

   _____   _____   _____   _____

   (3) Common oxides

   Al_2O_3 PbO N_2O CaO

   _____   _____   _____   _____

   (4) Common chlorides

   NH_4Cl NaCl   KCl $MgCl_2$

   _____   _____   _____   _____

5. Write the names for the formulas and the formulas for the written names listed below. Refer to the tables in Chapter 8.

   (1) AgCl _____

   (2) KNO_3 _____

   (3) $SnCl_4$ _____

   (4) Al_2O_3 _____

   (5) SO_3 _____

   (6) NH_4NO_3 _____

   (7) $AlPO_4$ _____

   (8) Na_2SO_4 _____

   (9) $Ca(OH)_2$ _____

   (10) $(NH_4)_2CO_3$ _____

   (11) NaI _____

(12) FeCl$_3$ _____

(13) BaO _____

(14) K$_2$Cr$_2$O$_7$ _____

(15) NaHCO$_3$ _____

(16) CO _____

(17) Hydrochloric acid _____

(18) Lead (II) nitrate _____

(19) Potassium permanganate _____

(20) Magnesium chloride _____

(21) Carbon dioxide _____

(22) Barium sulfate _____

(23) Iron (III) chloride _____

(24) Silicon dioxide _____

(25) Sulfuric acid _____

(26) Dinitrogen tetroxide _____

(27) Nitric acid _____

(28) Cobalt (II) chloride _____

(29) Sodium sulfite _____

(30) Aluminum hydroxide _____

(31) Hydrosulfuric acid _____

(32) Carbon tetrachloride _____

6. Give the formulas for the following salts, which contain more than one positive ion.

   (1) Sodium bicarbonate _____

   (2) Magnesium ammonium phosphate _____

   (3) Sodium potassium sulfate _____

   (4) Ammonium hydrogen sulfide _____

   (5) Potassium hydrogen sulfate _____

   (6) Potassium aluminum sulfate _____

7. Write the names for the following common hydroxides.

   (1) NaOH _____

   (2) KOH _____

   (3) Ca(OH)$_2$ _____

   (4) NH$_4$OH _____

8. Write the correct formulas for compounds formed by matching each cation with all anions.

	F^-	O^{2-}	NO_3^-	PO_4^{3-}
K^+				
Mg^{2+}				
Fe^{3+}				
H^+				
Cu^+				
Sn^{4+}				
Ca^{2+}				
Al^{3+}				

9. The following are selected rules for determining oxidation numbers of elements in compounds or ions.

Hydrogen is generally +1.
Oxygen is generally −2.

The algebraic sum of the oxidation numbers for all the atoms in a compound is equal to zero.

The algebraic sum of the oxidation numbers for all the atoms in a polyatomic ion is equal to the charge on the ion.

Determine the oxidation number for each indicated element in the following compounds or ions:

(1) S in H_2SO_3 _____
(2) N in HNO_3 _____
(3) Cr in $Cr_2O_7^{2-}$ _____
(4) Cl in ClO_4^- _____
(5) Zn in ZnO_2^{2-} _____
(6) Mn in MnO_4^- _____
(7) C in CH_4 _____
(8) C in $C_6H_{12}O_6$ _____
(9) Mn in $MnO(OH)_2$ _____
(10) C in H_2CO_3 _____
(11) C in $C_2O_4^{2-}$ _____
(12) As in AsO_4^{3-} _____

(13) P in P_2O_5 _____   (15) N in NO_2^- _____
(14) N in N_2O _____   (16) Br in BrO_3^- _____

10. Examine the following list of molecules and ions. Identify those that exhibit an ionic charge only, those with oxidation number values only, and those with both.

(1) Cl^-
(2) PO_4^{3-}
(3) BrCl
(4) Na^+
(5) H_2O
(6) Cl_2
(7) NH_4^+
(8) F_2

Challenge Problems

11. Determine the oxidation number of the indicated element in each of the following compounds.

(1) Cr in K_2CrO_4
(2) S in $Na_2S_2O_3$
(3) Ti in $Na_2Ti_3O_7$
(4) Pt in $H_2PtCl_6 \cdot 6H_2O$
(5) C in $KHO_4C_8H_4$
(6) Co in $Na_3Co(NO_2)_6$
(7) S in $(NH_4)_2S_2O_8$
(8) B in CaB_4O_7
(9) Mo in $(NH_4)_2MoO_4$
(10) As in $HAsO_3$

RECAP SECTION

Chapter 8 is one of the self-contained study units in the text for you to refer to as needed. After completing the material in the text, you will be familiar with naming binary compounds composed of a metal and a nonmetal and those composed of two nonmetals. Ternary compounds contain three elements. It is common to find oxygen-containing radicals in ternary compounds. Rules are given for naming ternary acids, bases, and salts. Only through practice and experience in a laboratory will you become adept at using chemical names. Try to use names whenever possible. Ask your instructor the name of any chemical you are not sure about. It is a good practice to have the original chemical bottles present during a lab period even though your instructor may provide the chemicals in solution. Ask to see the reagent bottles and read labels carefully.

ANSWERS TO QUESTIONS

1. (1) Potassium fluoride
Potassium oxide
Potassium nitrate
Potassium phosphate

    (2) Magnesium fluoride
Magnesium oxide
Magnesium nitrate
Magnesium phosphate

    (3) Iron (III) fluoride
Iron (III) oxide
Iron (III) nitrate
Iron (III) phosphate

    (4) Hydrogen fluoride
Water
Nitric acid
Phosphoric acid

    (5) Tin (IV) fluoride
Tin (IV) oxide
Tin (IV) nitrate
Tin (IV) phosphate

2. (1) Carbon monoxide
Carbon dioxide
Phosphorus pentachloride
Diphosphorus pentoxide

    (2) Carbon tetrachloride
Nitrogen oxide
Disulfur dichloride
Dinitrogen trioxide

3. (1) Ammonium bromide
Ammonium sulfide

    (2) Potassium hydroxide
Magnesium hydroxide

    (3) Sodium cyanide
Hydrogen cyanide

4. (1) Hydrochloric acid
Sulfuric acid
Acetic acid
Carbonic acid

    (2) Magnesium sulfate heptahydrate
Potassium carbonate
Sodium nitrate
Calcium phosphate

    (3) Aluminum oxide
Lead (II) oxide
Dinitrogen oxide
Calcium oxide

    (4) Ammonium chloride
Sodium chloride
Potassium chloride
Magnesium chloride

5. (1) Silver chloride
   (2) Potassium nitrate
   (3) Tin (IV) chloride
   (4) Aluminum oxide
   (5) Sulfur trioxide
   (6) Ammonium nitrate
   (7) Aluminum phosphate
   (8) Sodium sulfate
   (9) Calcium hydroxide
   (10) Ammonium carbonate
   (11) Sodium iodide
   (12) Iron (III) chloride
   (13) Barium oxide
   (14) Potassium dichromate
   (15) Sodium bicarbonate
   (16) Carbon monoxide
   (17) HCl
   (18) $Pb(NO_3)_2$
   (19) $KMnO_4$
   (20) $MgCl_2$
   (21) CO_2
   (22) $BaSO_4$
   (23) $FeCl_3$
   (24) SiO_2
   (25) H_2SO_4
   (26) N_2O_4
   (27) HNO_3
   (28) $CoCl_2$
   (29) Na_2SO_3
   (30) $Al(OH)_3$
   (31) H_2S
   (32) CCl_4

6. (1) $NaHCO_3$ (2) $MgNH_4PO_4$ (3) $NaKSO_4$
   (4) NH_4HS (5) $KHSO_4$ (6) $KAl(SO_4)_2$

7. (1) Sodium hydroxide   (2) Potassium hydroxide
   (3) Calcium hydroxide   (4) Ammonium hydroxide

8.
KF	K_2O	KNO_3	K_3PO_4
MgF_2	MgO	$Mg(NO_3)_2$	$Mg_3(PO_4)_2$
FeF_3	Fe_2O_3	$Fe(NO_3)_3$	$FePO_4$
HF	H_2O	HNO_3	H_3PO_4
CuF	Cu_2O	$CuNO_3$	Cu_3PO_4
SnF_4	SnO_2	$Sn(NO_3)_4$	$Sn_3(PO_4)_4$
CaF_2	CaO	$Ca(NO_3)_2$	$Ca_3(PO_4)_2$
AlF_3	Al_2O_3	$Al(NO_3)_3$	$AlPO_4$

9. (1) +4   (2) +5   (3) +6   (4) +7   (5) +2
   (6) +7   (7) −4   (8) 0   (9) +4   (10) +4
   (11) +3 (12) +5 (13) +5 (14) +1 (15) +3
   (16) +5

10. (1) ionic   (2) both   (3) oxidation number
    (4) ionic   (5) oxidation number   (6) oxidation number
    (7) both   (8) oxidation number

11. (1) +6   (2) +2   (3) +4   (4) +4   (5) +¼
    (6) +3   (7) +7   (8) +3   (9) +6   (10) +5

CHAPTER NINE
Quantitative Composition of Compounds

SELF-EVALUATION SECTION

1. In a chemical laboratory, substances are usually weighed out on a balance in grams or milligrams. It would be impossible to weigh out one atom or one molecule. Therefore, how do we relate particles on an atomic and molecular level to the relatively large quantity of mass called a gram? We do it very simply by stating that 1 atomic mass unit will become 1 gram when we wish to compare masses of chemicals and carry out laboratory experiments. If we express the atomic weight of an element in grams, we call this quantity 1 mole. The term *mole* is commonplace in chemistry and is used to describe a certain number of molecules, atoms, ions, or electrons. The actual number of particles contained in a mole of any substance is called (1) _____ _____ and has a value of (2) _____. It is possible to speak about a mole of bricks or a mole of moles, but the word is usually restricted to a chemical meaning. It is important to remember that a mole of a chemical compound not only refers to Avogadro's number of formula units but also relates to a corresponding mass of these formula units which can be weighed out in the laboratory.

2. Calculate the formula or molecular weights for the following compounds with the aid of the short list of atomic weights.

		Atomic Weights
(1) CCl_4 _____	Na	23.0
(2) $CaCO_3$ _____	C	12.0
(3) NH_4Cl _____	O	16.0
(4) $NaClO$ _____	Ca	40.1
	Cl	35.5
(5) $NaHCO_3$ _____	N	14.0
	H	1.0

Do your calculations here.

3. Using the molecular weights that were calculated for the second question, find the number of moles present for the following weights of chemical. Be sure the units cancel out properly after you set up the problem.

   (1) 77 g of CCl_4
       (mol wt = 154.0 g). Number of moles = _____

   (2) 33 g of $CaCO_3$
       (form. wt = 100.1 g). Number of moles = _____

   (3) 89 g of NH_4Cl
       (form. wt = 53.5 g). Number of moles = _____

   Do your calculations here.

4. Using the formula or molecular weights that were calculated for the second question, find the number of grams represented by the following number of moles of chemical. Be sure to cancel out the units to make sure you have set the problem up correctly.

   (1) 1.5 moles of NaClO (form. wt = 74.5 g)   _____ g
   (2) 0.67 mole of $NaHCO_3$ (form. wt = 84.0 g)   _____ g

   Do your calculations here.

5. Using the partial list of atomic weights, calculate the number of moles represented by a certain mass of an element.

    Na   23.0
    Ag   107.9
    S   32.1

(1) How many moles are represented by 46.0 g of Na atoms?

Do your calculations here.

(2) How many moles are represented by 54.0 g of Ag?

Do your calculations here.

(3) How many moles are represented by 100 g of S?

Do your calculations here.

6. Using the following list of atomic weights, calculate the elemental percentage composition for the four compounds listed.

Atomic weights: N = 14.0,   O = 16.0,   Al = 27.0,   C = 12.0,   Mn = 54.9, K = 39.1.

(1) N_2O (2) Al_2O_3 (3) K_2CO_3 (4) $KMnO_4$

Do your calculations here.

(1) N_2O

(2) Al_2O_3

(3) K_2CO_3

(4) $KMnO_4$

7. Fill in the blank space or circle the appropriate response.

The simplest formula of a compound, or the (1) _____ formula, tells us the (2) smallest/largest ratio of atoms present in a compound. The ratio is usually a small (3) _____ number ratio. It is possible for two compounds to have the same empirical formula, but different (4) _____ formulas. The true formula for a compound, or (5) _____ formula, represents the actual number of (6) _____ of each element found in one molecule of the compound. The weight of all the (7) _____ in a molecular formula is the compound's (8) _____ weight.

8. From the percentage composition data given for each compound, calculate the empirical formula.

(1) 86.6% Pb   Atomic wt. Pb = 207.2
    13.4% S   Atomic wt. S = 32.1

                            Empirical formula _____

Do your calculations here.

(2) 27.1% Na   Atomic wt. Na = 23.0
    16.5% N   Atomic wt. N = 14.0
    56.4% O   Atomic wt. O = 16.0

                            Empirical formula _____

Do your calculations here.

9. A certain compound is found to contain 1.45 g of Na, 2.05 g of S, and 1.5 g of O. With the aid of the list of atomic weights, calculate the empirical formula.

       Na Atomic wt. 23.0
       S   Atomic wt. 32.1
       O   Atomic wt. 16.0

                                      Formula _____

Do your calculations here.

10. The molecular weight of an unknown sugar substance is experimentally found to be 180 g/mole. The percentage composition data are determined to be 40% carbon, 7.0% hydrogen, and 53% oxygen. Calculate first the empirical formula and then the molecular formula.

       C   Atomic wt. 12.0
       H   Atomic wt. 1.0
       O   Atomic wt. 16.0

                                      Empirical _____

                                      Molecular _____

Do your calculations here.

11. A compound of bromine and iodine is formed by direct reaction between the two elements. It is found that 5.40 g of bromine react with 8.58 g of iodine. What is the empirical formula for the compound and if the molecular weight is 206.8, what is the molecular formula? What is the percentage composition of the compound?

Do your calculations here.

12. Calculate the number of grams and the number of atoms represented by the following quantities of two elements.

   (1) 1.5 moles of Si
       Atomic weight of Si = 28.1

   (2) 0.30 mole of Ca
       Atomic weight of Ca = 40.1

   Do your calculations here.

Challenge Problem

13. A compound known as cadaverine (1,5-Pentane diamine) is a ptomaine formed by the action of bacteria on meat and fish. Analysis shows that the elemental composition is C 58.8%, H 13.8% and N 27.4%. Determine the empirical formula and the molecular formula. ("Pentane" means 5 carbon atoms.)

Do your calculations here.

14. How many grams of Fe contain the same number of atoms as 154 g of arsenic?

    Do your calculations here.

15. At the present time the population of the United States is approximately 225,000,000. If one mole of pennies were distributed among the entire population, how many dollars would each person receive?

    Do your calculations here.

RECAP SECTION

Chapter 9 puts you to work using the periodic table and chemical formulas to try out your math skills on some basic chemical problem solving. You will be interested to know that, prior to the late 1890s, many chemists spent their careers working on problems of the elemental composition of various compounds. The techniques and mathematical steps that they used are all related to the problems that you have just solved. Even today it is necessary to determine the empirical and molecular formulas of the many new compounds synthesized each year.

ANSWERS TO QUESTIONS AND SOLUTIONS TO PROBLEMS

1. (1) Avogadro's number   (2) 6.022×10^{23}

2. (1) $12.0 + (4 \times 35.5) = 154.0 \dfrac{g}{mole}$

   (2) $40.1 + 12.0 + (3 \times 16.0) = 100.1 \dfrac{g}{mole}$

   (3) $14.0 + (4 \times 1.0) + 35.5 = 53.5 \dfrac{g}{mole}$

(4) $23.0 + 16.0 + 35.5 = 74.5 \dfrac{g}{mole}$

(5) $23.0 + 1.0 + 12.0 + (3 \times 16.0) = 84.0 \dfrac{g}{mole}$

3. (1) mol wt of CCl_4 = 154 g
Therefore,

$$77 \, g \times \dfrac{1 \text{ mole}}{154 \, g} = 0.50 \text{ mole}$$

(2) mol wt of $CaCO_3$ = 100.1 g
Therefore,

$$33 \, g \times \dfrac{1 \text{ mole}}{100.1 \, g} = 0.33 \text{ mole}$$

(3) mol wt of NH_4Cl = 53.5 g
Therefore,

$$89 \, g \times \dfrac{1 \text{ mole}}{53.5 \, g} = 1.7 \text{ moles}$$

4. (1) 1.5 moles of NaClO will equal

$$1.5 \text{ moles} \times 74.5 \, \dfrac{g}{mole} = 1.1 \times 10^2 \text{ g (2 sig. figures)}$$

(2) 0.67 mole of $NaHCO_3$ will equal

$$0.67 \text{ mole} \times 84.0 \, \dfrac{g}{mole} = 56 \text{ g (2 sig. figures)}$$

5. (1) 2.00 moles of Na
You are asked how many moles are contained in 46.0 g of Na atoms. One mole of any element is equal to the atomic weight expressed in grams. For Na this amount is 23.0 g. What we have said is that 23.0 g of Na equals 1.0 mole. We have 46.0 g, which is more than 1.0 mole. Arranging the problem so that the units will cancel out properly and give us an answer that is greater than one, we have

$$46.0 \, g \times \dfrac{1 \text{ mole}}{23.0 \text{ g}} = 2.00 \text{ moles}$$

(2) 0.500 mole
1 mole of Ag = 107.9 g
Therefore,

$$54.0 \, g \times \dfrac{1 \text{ mole}}{107.9 \, g} = 0.500 \text{ mole}$$

(3) 3.12 moles of S
The problem is solved just as the others have been.

$$100 \text{ g} \times \frac{1 \text{ mole}}{32.1 \text{ g}} = 3.12 \text{ moles of S}$$

6. (1) N_2O

$$\text{mol wt} = (2 \times 14.0) + 16.0 = 44.0 \ \frac{g}{\text{mole}}$$

Therefore,

$$\%N = \frac{2 \times 14.0}{44.0} \times 100\% = 63.6\%$$

$$\%O = \frac{16.0}{44.0} \times 100\% = 36.4\%$$

(2) Al_2O_3

$$\text{mol wt} = (2 \times 27.0) + (3 \times 16.0) = 102 \ \frac{g}{\text{mole}}$$

Therefore,

$$\%Al = \frac{2 \times 27.0}{102} \times 100\% = 52.9\%$$

$$\%O = \frac{3 \times 16.0}{102} \times 100\% = 47.1\%$$

(3) K_2CO_3

$$\text{mol wt} = (2 \times 39.1) + 12.0 + (3 \times 16.0) = 138.2 \ \frac{g}{\text{mole}}$$

Therefore,

$$\%K = \frac{2 \times 39.1}{138.2} \times 100\% = 56.6\%$$

$$\%C = \frac{12.0}{138.2} \times 100\% = 8.68\%$$

$$\%O = \frac{3 \times 16.0}{138.2} \times 100\% = 34.7\%$$

(4) $KMnO_4$

$$\text{mol wt} = 39.1 + 54.9 + (4 \times 16.0) = 158 \ \frac{g}{\text{mole}}$$

Therefore,

$$\%K = \frac{39.1}{158} \times 100\% = 24.7\%$$

$$\%Mn = \frac{54.9}{158} \times 100\% = 34.7\%$$

$$\%O = \frac{4 \times 16.0}{158} \times 100\% = 40.5\%$$

7. (1) empirical (2) smallest (3) whole- (4) molecular
   (5) molecular (6) atoms (7) atoms (8) molecular

8. (1) PbS (2) $NaNO_3$

   (1) Taking 100 g of the compound composed of Pb and S, we would have 87 g Pb and 13 g S. We need to determine the number of moles of Pb and S present in the 100 g. Therefore,

   $$Pb \quad 87\,g \times \frac{1 \text{ mole}}{207.2\,g} = 0.42 \text{ mole}$$

   $$S \quad 13\,g \times \frac{1 \text{ mole}}{32.1\,g} = 0.40 \text{ mole}$$

   The ratio of Pb to S is 0.42 to 0.40 or 1:1. The formula is PbS.

   (2) Likewise, in a 100 g of compound 2, we would have 27.1 g of Na, 16.5 g of N, and 56.4 g of O. The number of moles of each element would be:

   $$Na \quad 27.1\,g \times \frac{1 \text{ mole}}{23.0\,g} = 1.18 \text{ moles}$$

   $$N \quad 16.5\,g \times \frac{1 \text{ mole}}{14.0\,g} = 1.18 \text{ moles}$$

   $$O \quad 56.4\,g \times \frac{1 \text{ mole}}{16.0\,g} = 3.53 \text{ moles}$$

   To eliminate the decimals from our ratio we must divide each of the numbers by the smallest number.

   $$Na = \frac{1.18}{1.18} = 1 \qquad O = \frac{3.53}{1.18} = 2.99$$

   $$N = \frac{1.18}{1.18} = 1$$

   When the ratio is expressed as small whole numbers, the correct empirical formula becomes $NaNO_3$.

9. $Na_2S_2O_3$

We must determine the number of moles of each element present and then find the smallest whole-number ratio.

		Smallest Ratio
Na	$1.45 \text{ g} \times \dfrac{1 \text{ mole}}{23.0 \text{ g}} = 0.063$ mole	$\dfrac{0.063}{0.063} = 1$
S	$2.05 \text{ g} \times \dfrac{1 \text{ mole}}{32.1 \text{ g}} = 0.064$ mole	$\dfrac{0.064}{0.063} = 1$
O	$1.5 \text{ g} \times \dfrac{1 \text{ mole}}{16.0 \text{ g}} = 0.094$ mole	$\dfrac{0.094}{0.063} = 1.5$

Therefore, $1:1:1.5 = 2:2:3$
The empirical formula is $Na_2S_2O_3$

10. Empirical formula: CH_2O Molecular formula: $C_6H_{12}O_6$

The first step is to find the empirical formula from the number of moles of each element for a hypothetical 100 g of compound.

		Smallest Ratio
C	$40 \text{ g} \times \dfrac{1 \text{ mole}}{12.0 \text{ g}} = 3.3$ moles	$\dfrac{3.3}{3.3} = 1$
H	$7.0 \text{ g} \times \dfrac{1 \text{ mole}}{1.0 \text{ g}} = 7.0$ moles	$\dfrac{7.0}{3.3} = 2.1$
O	$53 \text{ g} \times \dfrac{1 \text{ mole}}{16.0 \text{ g}} = 3.3$ moles	$\dfrac{3.3}{3.3} = 1$

The rounded-off whole-number ratio would be $(CH_2O)_n$ and the empirical formula would therefore be CH_2O.

The molecular formula is calculated from the total atomic weights in the empirical formula and the molecular weight. The total weight of the empirical formula is $12.0 + 2.0 + 16.0 = 30.0$. Next, determine the ratio between the given molecular weight and the empirical formula weight.

$$\dfrac{180}{30} = 6$$

Therefore, we need to multiply the empirical formula by 6 to obtain the molecular formula.

$6 \times CH_2O$ would then be $C_6H_{12}O_6$

11. Empirical formula and molecular formula are the same, BrI. The percentage composition is 38.6% Br_2 and 61.3% I_2.

$$\text{Number of moles of } Br_2 = \dfrac{5.40 \text{ g}}{79.9 \text{ g/mole}} = 0.0676 \text{ moles}$$

Number of moles of $I_2 = \dfrac{8.58 \text{ g}}{126.9 \text{ g/mole}} = 0.0676$ moles

The number of moles of each element are in a ratio of 1:1, so the empirical formula is BrI. One mole of BrI weighs 206.8g, which matches the value given in the problem, so the molecular formula is also BrI. Percentage composition is calculated as follows:

$$\% \text{ Br} = \dfrac{\text{wt. Br}_2}{\text{total wt.}} = \dfrac{5.40 \text{ g}}{5.40 + 8.58 \text{ g}} \times 100\% = 38.6\%$$

Likewise

$$\% \text{ I}_2 = \dfrac{8.58 \text{ g}}{13.98 \text{ g}} \times 100\% = 61.4\%$$

12. (1) 42 g of Si and 9.0×10^{23} atoms Si

    Silicon has 28.1 g in 1 mole. We have 1.5 moles, which will be more than 28.1 g. Arranging the problem so that the units cancel out, we have:

    $$1.5 \text{ moles} \times 28.1 \dfrac{\text{g}}{\text{mole}} = 42 \text{ g of Si}$$

    To find the number of atoms in a certain number of moles, we simply multiply the number of moles by Avogadro's number.

    $$1.5 \text{ moles} \times 6.022 \times 10^{23} \dfrac{\text{atoms}}{\text{mole}} = 9.0 \times 10^{23} \text{ atoms Si}$$

    (2) 12 g of Ca and 1.8×10^{23} atoms Ca

    $$0.30 \text{ mole} \times 40.1 \dfrac{\text{g}}{\text{mole}} = 12 \text{ g of Ca}$$

    $$0.30 \text{ mole} \times 6.022 \times 10^{23} \dfrac{\text{atoms}}{\text{mole}} = 1.8 \times 10^{23} \text{ atoms Ca}$$

13. Empirical formula and molecular formula both $C_5H_{14}N_2$

    $$\text{moles of C} = \dfrac{58.8}{12.0} = 4.90 \qquad \text{moles of H} = \dfrac{13.8}{1} = 13.8$$

    $$\text{moles of N} = \dfrac{27.4}{14.0} = 1.96$$

    lowest common multiple is $C_5H_{14}N_2$

14. 115 g Fe. We can start by finding out how many moles of arsenic are represented by 154 g.

    $$\dfrac{154 \text{ g}}{74.9 \text{ g/mole}} = 2.06 \text{ mole As}$$

2.06 moles of As contains the same number of atoms as 2.06 moles of Fe. To convert this value into grams of Fe, we need only to multiply the atomic weight by the number of moles.

$$2.06 \text{ moles} \times 55.8 \frac{\text{g Fe}}{\text{mole}} = 115 \text{ g Fe (3 sig. figures)}$$

15. $\$2.68 \times 10^{13}$ per person. One mole of pennies equals one Avogadro's number of pennies or 6.022×10^{23} pennies. To find dollars, divide this number by 10^2 or 100.

$$\frac{6.022 \times 10^{23} \text{ pennies}}{1 \times 10^2 \frac{\text{pennies}}{\text{dollar}}} = 6.022 \times 10^{21} \text{ dollars}$$

$$\frac{6.022 \times 10^{21} \text{ dollars}}{2.25 \times 10^8 \text{ persons}} = \$2.68 \times 10^{13}/\text{person}$$

CHAPTER TEN
Chemical Equations

SELF-EVALUATION SECTION

1. Match the symbols used in chemical equations with the corresponding descriptive statements.

Symbols	
$\rightarrow$	(s)
$+$	(l)
$\rightleftarrows$	(g)
$\uparrow$	(Δ)
$\downarrow$	(aq)

    (1) Gas (written after substance) _____

    (2) Reversible reaction; equilibrium between reactants and products _____

    (3) Precipitate (written after substance) _____

    (4) Heat _____

    (5) Added to _____

    (6) Liquid (written after substance) _____

    (7) Aqueous solution (substance dissolved in water) _____

    (8) Yields; produces (points to products) _____

    (9) Solid (written after substance) _____

2. Balance the following equations.

    (1) $Fe + H_2O \rightarrow Fe_3O_4 + H_2$

(2) $H_2O_2 \rightarrow H_2O + O_2$

(3) $NH_4NO_2 \rightarrow N_2 + H_2O$

(4) $C_6H_{14} + O_2 \rightarrow CO_2 + H_2O$

(5) $CO + Fe_3O_4 \rightarrow FeO + CO_2$

Translate the word equations into formulas and balance them.

(6) Potassium nitrate → Potassium nitrite + Oxygen

(7) Calcium oxide + Hydrochloric acid → Calcium chloride + Water

(8) Copper metal + Sulfuric acid → Copper(II) sulfate + Water + Sulfur dioxide

(9) Bromine + Hydrogen sulfide → Hydrogen bromide + Sulfur

(10) Zinc sulfide + Oxygen → Zinc oxide + Sulfur dioxide

3. Balance the following equations.

   (1) K + H_2O → KOH + H_2

   (2) Ca + O_2 → CaO

   (3) Na_2O + H_2O → NaOH

   (4) Zn + HCl → $ZnCl_2$ + H_2

   (5) N_2 + H_2 → NH_3

Translate the word equations into formulas and balance them.

(6) Carbon + Oxygen → Carbon monoxide

(7) Sodium bicarbonate + Sulfuric acid → Sodium sulfate + Water + Carbon dioxide

(8) Lithium hydroxide + Hydrobromic acid → Lithium bromide + Water

(9) Potassium chromate + Lead(II) nitrate → Potassium nitrate + Lead(II) chromate

(10) Nitric acid + Calcium hydroxide → Calcium nitrate + Water

4. Identify the following reactions as combination (C), decomposition (D), single displacement (SD), or double displacement (DD).

(1) $2\ Al(OH)_3 + 3\ H_2SO_4 \rightarrow Al_2(SO_4)_3 + 6\ H_2O$ _____

(2) $4\ K + O_2 \rightarrow 2\ K_2O$ _____

(3) $Cl_2 + 2\ NaBr \rightarrow Br_2 + 2\ NaCl$ _____

(4) $2\ HgO \rightarrow 2\ Hg + O_2$ _____

(5) $MgCl_2 + 2\ AgNO_3 \rightarrow 2\ AgCl + Mg(NO_3)_2$ _____

(6) $CaO + H_2O \rightarrow Ca(OH)_2$ _____

(7) $2\ HCl + Na_2CO_3 \rightarrow 2\ NaCl + H_2O + CO_2$ _____

(8) $2\ KClO_3 + \rightarrow 2\ KCl + 3\ O_2$ _____

5. Identify the following reactions as exothermic (ex) or endothermic (en).
   (1) $N_2(g) + O_2(g) + 181 \text{ kJ} \rightarrow 2 NO(g)$ _____
   (2) $C(s) + O_2(g) \rightarrow CO_2(g) + 94.0 \text{ kcal}$ _____
   (3) $C_3H_8(g) + 5O_2(g) \rightarrow 3CO_2(g) + 4H_2O(g) + 2200 \text{ kJ}$ _____

6. Interpret the odd-numbered reactions of question 4 in terms of number of moles of reactants and products involved. Write your answers below.
   (1)

   (3)

   (5)

   (7)

Challenge Problem

7. Identify each of the following reactions as combination, decomposition, single displacement or double displacement. Complete and balance each one.
   (1) $KClO_3 \xrightarrow{\Delta}$
   (2) $Na + H_2O \longrightarrow$
   (3) $HCl + K_2CO_3 \longrightarrow$
   (4) $H_2 + N_2 \xrightarrow{\Delta}$
   (5) $CaCO_3 \xrightarrow{\Delta}$
   (6) $Zn + Pb(NO_3)_2 \longrightarrow$
   (7) $Al_2(SO_4)_3 + NH_4OH \longrightarrow$
   (8) $CaO_2 + H_2O \longrightarrow$
   (9) $Cl_2 + KBr \longrightarrow$
   (10) $Mg + Ni(Cl)_2 \longrightarrow$
   (11) $Ba(NO_3)_2 + Na_2SO_4 \longrightarrow$

(12) $NH_3 + HCl \longrightarrow$

(13) $H_2 + Cl_2 \longrightarrow$

(14) $HNO_3 + NaOH \longrightarrow$

(15) $NaCl + H_2SO_4 \longrightarrow$

(16) $Mg + H_2SO_4 \longrightarrow$

(17) $Na_2O + H_2SO_4 \longrightarrow$

(18) $Mg + O_2 \xrightarrow{\Delta}$

(19) $NaI + Cl_2 \longrightarrow$

(20) $AgNO_3 + NaCl \longrightarrow$

8. Will a reaction occur when the following are mixed? If so, write a balanced equation. Use the short activity series in the chapter.

   (1) $Al_{(s)} + NaBr_{(aq)}$
   (2) $Zn_{(s)} + CuCl_{2(aq)}$
   (3) $Cu_{(s)} + HCl_{(aq)}$
   (4) $Mg_{(s)} + NiCl_{2(aq)}$
   (5) $K_{(s)} + H_2O$

RECAP SECTION

Chapter 10 is a self-contained section similar to Chapter 8. You should refer to both of these chapters from time to time to review nomenclature and chemical equations. You have learned what a chemical equation is and how to put one together in a balanced form. We are now able to use a chemical equation rather than a word equation to describe chemical changes.

ANSWERS TO QUESTIONS

1. (1) ↑ or (g)   (2) ⇌   (3) ↓   (4) Δ   (5) +
   (6) (l)   (7) (aq)   (8) →   (9) (s)

2. (1) $3 Fe + 4 H_2O \rightarrow Fe_3O_4 + 4 H_2$
   (2) $2 H_2O_2 \rightarrow 2 H_2O + O_2$
   (3) $NH_4NO_2 \rightarrow N_2 + 2 H_2O$
   (4) $2 C_6H_{14} + 19 O_2 \rightarrow 12 CO_2 + 14 H_2O$
   (5) $CO + Fe_3O_4 \rightarrow 3 FeO + CO_2$
   (6) $2 KNO_3 \rightarrow 2 KNO_2 + O_2$
   (7) $CaO + 2 HCl \rightarrow CaCl_2 + H_2O$
   (8) $Cu + 2 H_2SO_4 \rightarrow CuSO_4 + 2 H_2O + SO_2$
   (9) $Br_2 + H_2S \rightarrow 2 HBr + S$
   (10) $2 ZnS + 3 O_2 \rightarrow 2 ZnO + 2 SO_2$

3. (1) $2K + 2H_2O \rightarrow 2KOH + H_2$
   (2) $2Ca + O_2 \rightarrow 2CaO$
   (3) $Na_2O + H_2O \rightarrow 2NaOH$
   (4) $Zn + 2HCl \rightarrow ZnCl_2 + H_2$
   (5) $N_2 + 3H_2 \rightarrow 2NH_3$
   (6) $2C + O_2 \rightarrow 2CO$
   (7) $2NaHCO_3 + H_2SO_4 \rightarrow Na_2SO_4 + 2H_2O + 2CO_2$
   (8) $LiOH + HBr \rightarrow LiBr + H_2O$
   (9) $K_2CrO_4 + Pb(NO_3)_2 \rightarrow 2KNO_3 + PbCrO_4$
   (10) $2HNO_3 + Ca(OH)_2 \rightarrow Ca(NO_3)_2 + 2H_2O$

4. (1) DD   (2) C   (3) SD   (4) D   (5) DD
   (6) C   (7) DD   (8) D

5. (1) en   (2) ex   (3) ex

6. (1) Reactants 2 moles $Al(OH)_3$ and 3 moles H_2SO_4
       Products   1 mole $Al_2(SO_4)_3$ and 6 moles H_2O

   (3) Reactants 1 mole Cl_2 and 2 moles NaBr
       Products   1 mole Br_2 and 2 moles NaCl

   (5) Reactants 1 mole $MgCl_2$ and 2 moles $AgNO_3$
       Products   2 moles AgCl and 1 mole $Mg(NO_3)_2$

   (7) Reactants 2 moles HCl and 1 mole Na_2CO_3
       Products   2 moles NaCl, 1 mole H_2O and 1 mole CO_2

7. (1) decomposition $2KClO_3 \xrightarrow{\Delta} 2KCl + 3O_2 \uparrow$

   (2) single displacement $2Na + 2H_2O \longrightarrow 2NaOH + H_2 \uparrow$

   (3) double displacement $2HCl + K_2CO_3 \longrightarrow 2KCl + H_2O + CO_2 \uparrow$

   (4) combination $3H_2 + N_2 \xrightarrow{\Delta} 2NH_3$

   (5) decomposition $CaCO_3 \xrightarrow{\Delta} CaO + CO_2 \uparrow$

   (6) single displacement $Zn + Pb(NO_3)_2 \longrightarrow Zn(NO_3)_2 + Pb$

   (7) double displacement $Al_2(SO_4)_3 + 6NH_4OH \longrightarrow 3(NH_4)_2SO_4 + 2Al(OH)_3$

   (8) combination $CaO + H_2O \longrightarrow Ca(OH)_2$

   (9) single displacement $Cl_2 + 2KBr \longrightarrow Br_2 + 2KCl$

   (10) single displacement $Mg + Ni(Cl)_2 \longrightarrow Ni + Mg(Cl)_2$

   (11) double displacement $Ba(NO_3)_2 + Na_2SO_4 \longrightarrow BaSO_4 \downarrow + 2NaNO_3$

   (12) combination $NH_3 + HCl \longrightarrow NH_4Cl$

(13) combination　　$H_2 + Cl_2 \longrightarrow 2HCl$

(14) double displacement　　$HNO_3 + NaOH \longrightarrow NaNO_3 + H_2O$

(15) double displacement　　$NaCl + H_2SO_4 \longrightarrow NaHSO_4 + HCl\uparrow$

(16) single displacement　　$Mg + H_2SO_4 \longrightarrow MgSO_4 + H_2\uparrow$

(17) double displacement　　$Na_2O + H_2SO_4 \longrightarrow Na_2SO_4 + H_2O$

(18) combination　　$2Mg + O_2 \xrightarrow{\Delta} 2MgO$

(19) single displacement　　$2NaI + Cl_2 \longrightarrow 2NaCl + I_2$

(20) double displacement　　$AgNO_3 + NaCl \longrightarrow AgCl\downarrow + NaNO_3$

8. (1) No

(2) Yes　$Zn_{(s)} + CuCl_{2(aq)} \longrightarrow Cu_{(s)} + ZnCl_{2(aq)}$

(3) No

(4) Yes　$Mg_{(s)} + NiCl_{2(aq)} \longrightarrow Ni_{(s)} + MgCl_{2(aq)}$

(5) Yes　$2K + 2H_2O \longrightarrow 2KOH + H_{2(g)}\uparrow$

CHAPTER ELEVEN
Calculations from Chemical Equations

SELF-EVALUATION SECTION

1. Fill in with the appropriate response.

$$2\ Na + Cl_2 \rightarrow 2\ NaCl$$

In the above equation, (1) _____ mole(s) Na react with (2) _____ mole(s) Cl_2 to give (3) _____ mole(s) NaCl. The mole ratio of Na to Cl_2 is (4) _____. The mole ratio of Cl_2 to NaCl is (5) _____. If 7.0 moles of Na react, (6) _____ mole(s) of Cl_2 react with it to produce (7) _____ mole(s) of NaCl.

$$Pt + 8\ HCl + 2\ HNO_3 \rightarrow \underset{\text{Chloroplatinic Acid}}{H_2PtCl_6} + 2\ \underset{\text{Nitrosyl Chloride}}{NOCl} + 4\ H_2O$$

In the above equation, (8) _____ mole(s) of Pt reacts with (9) _____ mole(s) HCl and (10) _____ mole(s) HNO_3 to give (11) _____ mole(s) of H_2PtCl_6, (12) _____ mole(s) NOCl, and (13) _____ mole(s) H_2O. The mole ratio of Pt to HCl is (14) _____; the mole ratio of HCl to H_2PtCl_6 is (15) _____. If 0.300 mole of Pt is reacted, (16) _____ mole(s) HCl and (17) _____ mole(s) of HNO_3 will react with the Pt and (18) _____ mole(s) of H_2PtCl_6 or (19) _____ grams of H_2PtCl_6 will be produced.

Do your calculations here.

2. $2\ C_6H_{14} + 19\ O_2 \rightarrow 12\ CO_2 + 14\ H_2O$

In the above equation, the mole ratio of C_6H_{14} to CO_2 is (1) _____, and the mole ratio of C_6H_{14} to H_2O is (2) _____. If oxygen is present in abundance to carry out the reaction, which reagent will be the limiting reactant? (3) _____. How many moles of CO_2 can be produced from 3 moles of C_6H_{14}? (4) _____. If only 1.0 mole of C_6H_{14} is available for the reaction, what is the theoretical yield of CO_2 in moles and in grams? (5) _____; (6) _____. For the same amount of C_6H_{14} (1 mole), what is the theoretical yield of H_2O in moles and in grams? (7) _____; (8) _____.

After carrying out the above reaction with 1.0 mole of C_6H_{14}, a chemist measured an actual yield of 210 g of CO_2 and 115 g of water. What was his percentage yield of CO_2? (9) _____. His percentage yield of H_2O? (10) _____.

Do your calculations here.

3. We will now use the techniques of Chapter 11 to gain further chemical equation problem-solving skill.

Balance the following equation and then calculate the asked-for quantities.

$$PbO_2 \xrightarrow{\Delta} PbO + O_2 \uparrow$$

(1) How many grams of O_2 can be obtained from 100 grams of PbO_2? This is theoretical yield. Remember that we need to (a) use a balanced equation, (b) determine the number of moles of starting substance, (c) calculate the number of moles of desired substance using the mole-ratio technique, and (d) convert moles to grams of desired substance.

(2) What is the percentage yield if the actual yield of O_2 in the above reaction was 5.0 grams?

Do your calculations here.

4. Very often in industrial chemical processes, one of the reactants will be present in an amount that exceeds the requirements of the balanced equation. The reactant that is not in excess will therefore limit the amount of product that is formed and is named the limiting reactant. Using the equation given, answer the questions below.

$$2\ Al(OH)_3 + 3\ H_2SO_4 \rightarrow Al_2(SO_4)_3 + 6\ H_2O$$

(1) Reaction is run with excess of sulfuric acid. What is the limiting reactant?
(2) You have 9 moles of H_2SO_4 present for the reaction. What is the amount (moles) of $Al(OH)_3$ required to react completely with this amount of H_2SO_4?
(3) If 4 moles of $Al(OH)_3$ is the amount available, how much of the 9 moles of H_2SO_4 can be used?
(4) Using the 4 moles of $Al(OH)_3$, how many moles $Al_2(SO_4)_3$ and H_2O will you be able to produce?

Do your calculations here.

5. In the following reaction, how many moles of ZnO can be obtained? Also determine which reactant is the limiting reactant and which reactant is in excess.

$$2\ ZnS + 3\ O_2 \rightarrow 2\ ZnO + 2\ SO_2$$
100 g   100 g

Do your calculations here.

6. Balance the following equation and calculate how many moles of Cu can be formed from 5.0 moles of Al and 10 moles of $CuSO_4$. What is the limiting reactant and how much of the excess reactant is left after the reaction?

$$Al + CuSO_4 \rightarrow Cu + Al_2(SO_4)_3$$

5.0 mole   10.0 mole

Do your calculations here.

7. What is the theoretical yield of Fe_2O_3 that can be produced from 3.0 kg of Fe according to the following unbalanced equation?

$$Fe + O_2 \rightarrow Fe_2O_3$$

Do your calculations here.

Challenge Problems

8. It is possible to reclaim silver from used photographic fixer by using the active metal, powdered zinc. Zinc replaces the silver in solution, followed by conversion of the silver to silver oxide. After filtering, the silver oxide is reduced to metallic silver by carbon. The series of reactions are:

(1) $2AgBr + Zn \longrightarrow ZnBr_2 + 2Ag$

(2) $4Ag + O_2 \longrightarrow 2Ag_2O$

(3) $2Ag_2O + C \longrightarrow CO_2 + 4Ag$

How much silver can be reclaimed from 2.000 gallons of used fixer (density 1.018g/mL) if the silver concentration is 300 parts per million? Part per million is a general purpose concentration term that can take on a variety of units. For example, 1 µg per gram, 1 µL per liter and 1 mg per kilogram are examples of 1 ppm concentration.

Do your calculations here.

9. Refer to the problem above and the last reaction. The carbon that reduces the Ag_2O to metallic Ag comes from the filter paper that was used in the filtration step following Reaction (2). Assuming the filter paper to be pure cellulose with an empirical formula of $C_6H_{10}O_5$, what weight of filter paper is required to reduce the 2.31 g of silver contained in the two gallon volume of fixer?

Do your calculations here.

RECAP SECTION

The principal objective of Chapter 11 is to present a logical method for attacking problem solving associated with chemical equations. Since chemistry deals with chemical reactions and reactions are expressed in equation form, it follows that calculations associated with reactions are about as relevant and practical as any topic in chemistry. Professionals in agriculture, home economics, forestry, biology, and chemical engineering, in addition to chemists, are constantly working with stoichiometric calculations. (Stoichiometric calculations is the proper term to use for the material presented in Chapter 11.)

The technique of using the mole-ratio method for stoichiometric problems is very important to learn. Once you have a balanced equation, it is possible to set up any ratio between two species in the equation. If the problem asks for grams of reactant to produce so many grams of product, there will be additional calculation steps to perform on each side of the ratio, but the ratio is the connecting link.

grams reactant A → moles reactant A → moles product B → **grams product B**

ANSWERS TO QUESTIONS AND SOLUTIONS TO PROBLEMS

1. (1) 2   (2) 1   (3) 2   (4) 2:1   (5) 1:2
   (6) 3.5   (7) 7   (8) 1   (9) 8   (10) 2
   (11) 1   (12) 2   (13) 4   (14) 1:8   (15) 18:1
   (16) 2.40   (17) 0.600   (18) 0.300
   (19) 0.300 mole times the mol wt will give us the number of grams produced by the reaction. The mol wt of

   $$H_2PtCl_6 = (2 \times 1.0) + 195.1 + (6 \times 35.5) = 410.1 \; \frac{g}{mol}$$

   $$\text{Number of grams} = 410.1 \; \frac{g}{mol} \times 0.300 \; mol$$

   $$= 123 \; g \; H_2PtCl_6 \; (3 \; \text{sig. figures})$$

2. (1) 2:12 or 1:6   (2) 2:14 or 1:7   (3) C_6H_{14}
   (4) 18   (5) 6
   (6) 6 moles times the formula wt will give us the number of grams of CO_2 produced by the reaction.

   $$\text{The formula wt of } CO_2 = 12.0 + (2 \times 16.0) = 44.0 \; \frac{g}{mol}$$

   $$\text{Number of grams} = 44.0 \; \frac{g}{mol} \times 6 \; mol = 264 \; g \; CO_2$$

   (7) 7 moles
   (8) 7 moles times the formula wt will give us the number of grams of H_2O produced by the reaction.

   $$\text{The formula wt of } H_2O = (2 \times 1.0) + 16.0 = 18.0 \; \frac{g}{mol}$$

   $$\text{Number of grams} = 18.0 \; \frac{g}{mol} \times 7 \; mol = 126 \; g \; H_2O$$

   (9) Percentage yield of

   $$CO_2 = \frac{\text{actual yield}}{\text{theoretical yield}} \times 100\%$$

   $$= \frac{210 \; g}{264 \; g} \times 100\%$$

   $$= 79.5\%$$

   (10) Percentage yield of

   $$H_2O = \frac{\text{actual yield}}{\text{theoretical yield}} \times 100\%$$

   $$= \frac{115 \; g}{126 \; g} \times 100\%$$

   $$= 91.3\%$$

3. (1) 6.69 g of O_2

   The equation must be balanced before we can determine a proper mole ratio for calculating the amount of O_2.

   $$2\ PbO_2 \xrightarrow{\Delta} 2\ PbO + O_2 \uparrow$$

   Next, we need to convert 100 grams of PbO_2 into the number of moles of PbO_2.

   The formula wt of $PbO_2 = 207.2 + (2 \times 16.0) = 239.2\ \frac{g}{mol}$

   Number of moles of $PbO_2 = 100\ g \times \dfrac{1\ mol}{239.2\ g} = 0.418\ mol$

   The mole ratio is $\dfrac{mol\ desired\ substance}{mol\ starting\ substance} = \dfrac{1\ mol\ O_2}{2\ mol\ PbO_2}$

   moles of $O_2 = 0.418\ mol\ PbO_2 \times \dfrac{1\ mol\ O_2}{2\ mol\ PbO_2} = 0.209\ mol\ O_2$

   To convert moles of O_2 into grams, we multiply the number of moles by the molecular weight.

   $0.209\ mol \times 32.0\ \dfrac{g}{mol} = 6.69\ g\ O_2$

   (2) 75%

   The percentage yield is determined by dividing the actual yield by the theoretical yield multiplied by 100.

   $\dfrac{actual}{theoretical} \times 100\%$

   $\dfrac{5.0\ g}{6.7\ g} \times 100\% = 75\%$

4. (1) $Al(OH)_3$, aluminum hydroxide (2) 6 moles (3) 6 moles
   (4) 2 moles $Al_2(SO_4)_3$, 12 moles H_2O

5. 1.02 mole ZnO, ZnS is limiting reactant, O_2 is in excess. First we need to determine the number of moles of ZnO that can be obtained from each of the reactants.

   $100\ g\ ZnS \times \dfrac{1\ mol\ ZnS}{97.5\ g\ ZnS} \times \dfrac{2\ mol\ ZnO}{2\ mol\ ZnS} = 1.02\ mol\ ZnO$

   $100\ g\ O_2 \times \dfrac{1\ mol\ O_2}{32\ g\ O_2} \times \dfrac{2\ mol\ ZnO}{3\ mol\ O_2} = 2.08\ mol\ ZnO$

6. The balanced equation is

   $$2Al + 3CuSO_4 \longrightarrow 3Cu + Al_2(SO_4)_3$$

First we need to determine the number of moles of Cu that can be formed from each reactant.

$$5.0 \text{ mol Al} \times \frac{3 \text{ mol Cu}}{2 \text{ mol Al}} = 7.5 \text{ mol Cu}$$

$$10.0 \text{ mol CuSO}_4 \times \frac{3 \text{ mol Cu}}{3 \text{ mol CuSO}_4} = 10.0 \text{ mol Cu}$$

Therefore Al is the limiting reactant and 7.5 mol of Cu can be formed. Next we need to calculate the number of moles of $CuSO_4$ that will react with 5 moles of Al.

$$5 \text{ mol Al} \times \frac{3 \text{ mol CuSO}_4}{2 \text{ mol Al}} = 7.5 \text{ mol CuSO}_4$$

Therefore, 10.0 mol $CuSO_4$ − 7.5 mol $CuSO_4$ = 2.5 mol of $CuSO_4$ in excess.

7. 4300 g Fe_2O_3 (3 sig. figures)
   The balanced equation is

$$4 \text{ Fe} + 3 \text{ O}_2 \rightarrow 2 \text{ Fe}_2\text{O}_3$$

The number of moles of Fe = $3.0 \text{ kg} \times 1000 \frac{\text{g}}{\text{kg}} \times \frac{1 \text{ mol}}{55.8 \text{ g}}$

= 53.8 mol of Fe

moles of Fe_2O_3 = $53.8 \text{ mol Fe} \times \frac{2 \text{ mol Fe}_2\text{O}_3}{4 \text{ mol Fe}}$ = 26.9 mol

To determine the number of grams, we multiply 26.9 moles of Fe_2O_3 by the formula weight.

The formula wt of Fe_2O_3 = (2 × 55.8) + (3 × 16.0) = 160 $\frac{\text{g}}{\text{mol}}$

$$26.9 \text{ mol} \times 160 \frac{\text{g}}{\text{mol}} = 4300 \text{ g Fe}_2\text{O}_3 \text{ (3 sig. figures)}$$

8. 2.31 g
   Two gallons is equal to 7568 mL which has a density of 1.018 g/mL. The mass of the solution therefore is:

$$(1.018 \text{ g /mL}) (7568 \text{ mL}) = 7704 \text{ g}$$

Of this mass, 7704 g, 300 ppm are silver ion. To calculate how many grams of silver might be reclaimed with no loss, we set up the following equation.

$$\left(\frac{1}{1 \times 10^6 \text{ ppm}}\right)(300 \text{ ppm}) (7704 \text{ g}) = 2.31 \text{ g (3 sig. figures)}$$

9. 0.144 g filter paper
   The equation tells us that 1 mole of C reduces 4 moles of Ag ion to metallic silver.

Therefore, moles of carbon required is equal to

$$\frac{2.31 \text{ g Ag}}{107.9 \text{ g/mol Ag}} \times \frac{1 \text{ mol C}}{4 \text{ mol Ag}} = 0.00535 \text{ mol carbon}$$

We need a piece of filter paper that will contain at least 0.00535 moles carbon. The percentage of carbon in the cellulose is

$$\%C = \frac{72}{162} \times 100 = 44.4\%$$

To calculate the weight of filter paper needed we can convert the number of moles of carbon into grams and divide this value by the amount of carbon in the paper.

$$\text{wt. paper} = (0.00535 \text{ mol C})(12.0 \text{ g/mol})\left(\frac{1}{0.444}\right) = 0.144 \text{ g}$$

CHAPTER TWELVE
The Gaseous State of Matter

SELF-EVALUATION SECTION

1. The science of chemistry on a quantitative basis began with the systematic study of gas behavior, which may seem somewhat ironic since substances in the gas state are difficult to handle and use for experimental purposes. The fundamental properties of gases were established as compressibility, diffusion, pressure, and expansion. In addition, equal volumes of gases at the same temperature were found to exert identical pressures.

   The assumptions of the kinetic-molecular theory for an ideal gas are listed below.
   a. Gases consist of tiny particles.
   b. The volume of gas is mostly empty space.
   c. Gas molecules have no attraction for each other.
   d. Gas molecules move in straight lines in all directions, undergoing frequent collisions.
   e. No energy is lost through the collisions of gas molecules.
   f. The average kinetic energy for molecules is the same for all gases at the same temperature.

   Properties—list the letter for one or more assumptions from the above list that describe each of the following properties of gases.

   (1) Compressibility   _____

   (2) Diffusion   _____

   (3) Pressure   _____

   (4) Expansion   _____

   (5) Pressure of equal volumes of gases _____

2. An important assumption of the kinetic-molecular theory is that gas molecules move in straight lines and collide with each other and the walls of the container. Pressure, whether it's water pressure or gas pressure, is the force exerted on a unit area by a substance. As gas molecules collide with the container walls, they exert a certain pressure. What will happen to the pressure of a gas sample

if more gas molecules are placed in a container? More molecules mean more collisions with the walls. Therefore, the pressure will rise. If the temperature and volume of the container are kept constant, there is a direct relationship between pressure and number of gas molecules. Thus, doubling the amount of gas will double the pressure. Reducing the amount of gas to one-third the original quantity will reduce the pressure correspondingly. Refer to Figure 12.4 in the text.

3. Boyle's law and Charles' law.
What happens to the volume of a gas when the pressure is increased? Is the volume increased or reduced? We know from experience that gases can be compressed so that an increase in pressure will reduce the volume, but is there a quantitative or mathematical relationship that might be useful? Robert Boyle showed that, at constant temperature (T), the volume (V) of a gas is inversely proportional to the pressure (P). Thus, increased pressure reduces the volume, and reduced pressure results in increased volume. There are different ways to present Boyle's law, and you will be asked to choose the correct ways in a moment. Before we do that, what happens to a volume of gas when the absolute temperature increases? We know gases have kinetic energy. Since the mass of the gas molecules is constant, the velocity must increase. The question now is: Will the gas expand to a new volume? Yes, it will if the pressure of the gas can be maintained at a constant value. Thus, at constant pressure the volume of a gas is directly proportional to the absolute temperature. If the temperature goes up, the volume goes up, and vice versa. What happens to a gas sample if the pressure is not constant and the volume cannot expand? Visualize a closed empty can placed on a fire. The temperature of the gas increases; kinetic energy and pressure increase. The gas cannot expand so the pressure increases to the point at which the mechanical strength of the can cannot withstand the high pressure, and the can explodes.

For our study of gas behavior, we will be concerned only with constant-pressure systems as described in Charles' law.

Given below are several ways of mathematically stating Boyle's law and Charles' law. Place a "B" (for Boyle's) or "C" (for Charles') by each formula.

(1) $PV = k$ _____

(2) $\dfrac{V_1}{T_1} = \dfrac{V_2}{T_2}$ _____

(3) $V \propto \dfrac{1}{P}$ _____

(4) $P_1 V_1 = P_2 V_2$ _____

(5) $V \propto T$ _____

(6) $\dfrac{V}{T} = k$ _____

The symbol "$\propto$" means "to vary" or "to be proportional to."

4. Which of the gases in each pair will effuse at the fastest rate according to Graham's law?

(1) $N_2 + F_2$

(2) $CO_2 + Ne$

(3) $SO_2 + Br_2$

(4) $H_2 + He$

(5) $Cl_2 + O_2$

(6) $NH_3 + CO$

5. Gas law problems.

   (1) The pressure of a gas in a piston is 2100 torr. The pressure is reduced to 550 torr. The original volume was 1.00 liter; what is the new volume?

   Do your calculations here.

   (2) A certain gas exerted a pressure of 785 torr in a 3.00-liter container. The gas was compressed into a 250-mL gas bottle. What was the new pressure?

   Do your calculations here.

   (3) 100 mL of gas are obtained from a reaction at 300°C. What volume will the gas occupy at 100°C?

   Do your calculations here.

(4) A 250-mL sample of gas originally at $-50°C$ was brought to room temperature ($25°C$). Find the new volume.

Do your calculations here.

(5) You should also be able to work a combined gas law problem that involves P, V, and T. Try this one.

What is the volume of a gas at STP if the original conditions were 45 mL, 0.75 atm pressure, and $21°C$?

Do your calculations here.

6. Fill in the blank or circle the appropriate response.

We have learned in Chapter 12 that at standard temperature and pressure, which are (1) _____ and (2) _____, 1 mole of any gas occupies (3) _____ liters. This volume is known as the gram-molecular volume and is constant for any gas at STP. Another way to state the relationship is to say that Avogadro's Number of gas molecules (6.022×10^{23}) at STP is equal to 1 molar volume. When performing calculations involving gases, the usual practice is to convert all volumes of gases to standard conditions. Then the relationship 1 mole = 22.4 liters can be used for stoichiometric calculations. As a summary statement about gas behavior, one might state Avogadro's Law, which says that equal volumes of gases at the (4) same/different temperature and pressure contain equal numbers of molecules. This statement follows from the assumption that two gases at the same temperature possess the same average kinetic energy. If they occupy the same volume, the pressures will be the same. This is a consequence of the same number of molecules moving with the same average kinetic energy.

7. The density of a gas is expressed in grams per liter.

$$d = \frac{\text{mass}}{\text{volume}} = \frac{g}{L}$$

At STP, density may also be calculated from the following equation.

$$\text{density at STP} = \frac{\text{formula wt}}{22.4 \text{ L/mol}} = \frac{\text{g/mol}}{\text{liters/mol}} = \frac{g}{L}$$

(1) At STP, what is the density of NO (nitrogen oxide) gas? Use your table of atomic weights.

Do your calculations here.

(2) The density of HCN (hydrogen cyanide) gas at STP is 1.21 g/L. What volume will 100 g of HCN occupy at STP?

Do your calculations here.

(3) A quick way to prepare acetylene gas in the laboratory is to drop chunks of calcium carbide into water. Old-fashioned miners' lamps used this reaction, as did some residential lighting systems years ago. We will work a Dalton's partial pressure problem from this reaction.

A sample of C_2H_2 (acetylene) gas collected over water at 21°C and 750 torr pressure occupies a volume of 175 mL. Calculate the volume of dry acetylene at STP. The vapor pressure of water at various temperatures is listed in Appendix II of the textbook. Dalton's law states that, in a mixture of gases, each gas exerts its own individual pressure.

$$P_{total} = P_A + P_B + P_C$$

For this particular problem, the total pressure of the moist acetylene collected is made up of two parts—the pressure of C_2H_2 and the pressure of the water vapor. We have to subtract the partial pressure of the water vapor from the C_2H_2 pressure before we correct the C_2H_2 volume to STP.

Do your calculations here.

8. What volume of NO gas will be produced from 4.00 moles of N_2 and 3.00 moles of O_2 at STP according to the following equation?

$$N_{2(g)} + O_{2(g)} \longrightarrow 2\ NO_{(g)}$$

Do your calculations here.

9. What volume of NO at STP gas can be produced from 7.0 moles of nitrogen dioxide (NO_2) according to the following equation?

$$3\ NO_2 + H_2O \rightarrow 2\ HNO_3 + NO$$

Do your calculations here.

10. Using the ideal gas equation, calculate the volume that 15 g of H_2 gas at 25°C and 1.2 atm pressure will occupy.

$$PV = nRT$$

Do your calculations here.

Challenge Problems

11. In a movie thriller, the hero is locked in a sealed room. The bad guys allow a mixture of two poisonous gases, arsine (AsH_3—garlic odor) and cyanogen (C_2N_2—almond odor), to effuse through a porous opening into the room. If the hero doesn't find an escape route, what will be the fragrance that reaches the nostrils first?

12. The size cylinder known as 1A cylinder has a volume of 43.8 L. How many moles of oxygen are contained in a 1A cylinder at 72° F and 1500 pounds per in^2 pressure? To convert psi to atmospheres multiply psi by 0.06805 atm/psi. What is the mass of this number of moles of oxygen gas?

Do your calculations here.

13. Use the ideal gas equation to calculate the molecular weight of arsine gas (the garlic odor from problem 10), given the following information.

3.48 g of AsH_3 at 740 torr and 21°C occupies 1.11 L of volume. What is the molecular weight of AsH_3?

Do your calculations here.

RECAP SECTION

Chapter 12 is long and contains a great deal of new material. However, you should have sharpened up your problem-solving skills and learned a great deal about the gaseous state. Particularly important is the ability to visualize in equation form a word statement such as Boyle's law and then use the relationship to solve problems. The relationships of observed gas properties to the kinetic-molecular theory of ideal gas behavior, Avogadro's Law, Gay-Lussac's law of combining volumes, and Dalton's Law of Partial Pressures were studied. However, it is important to realize that real gases don't exactly behave as the ideal gas equation predicts because real gases do have a finite volume and also exhibit intermolecular attractions, particularly at low temperatures and high pressures. You should feel confident in tackling almost any stoichiometric problem, given an equation and an atomic weight table. You will use these skills again in later chapters on solution chemistry.

ANSWERS TO QUESTIONS AND SOLUTIONS TO PROBLEMS

1. (1) Compressibility—b
   (2) Diffusion—c, d
   (3) Pressure—d
   (4) Expansion—b, c, d
   (5) Pressure of equal volumes of gases—f, e

2. Assumption "a" is based on the experiments which established the relative size of any atom regardless of its state (solid, liquid, or gas). Assumption "e" is verified by experiments that show no temperature loss for gas samples upon standing.

3. (1) B (2) C (3) B (4) B (5) C (6) C

4. (1) N_2 (2) Ne (3) SO_2 (4) H_2 (5) O_2 (6) NH_3

5. (1) 3.82 liters

The problem involves pressures and volumes; this means a Boyle's law problem.

The original volume was 1.00 L, and the original pressure was 2100 torr. The pressure dropped to 550 torr. What happens to the volume of a gas when the pressure goes down? According to Boyle's law, the volume increases. This means we must multiply the original volume by a ratio of pressures, which will give us a larger volume than 1.00 L.

original volume × ratio of pressure = new volume

$$1.00 \text{ L} \times \frac{2100 \text{ torr}}{550 \text{ torr}} = 3.82 \text{ L}$$

(2) 9.42×10^3 torr

The problem again deals with pressures and volumes, but the two volumes do not have the same units. After making this conversion, we ask how the volume change will affect the pressure. The gas is being compressed into a smaller volume. The pressure will then increase; therefore the ratio must be greater than one.

original pressure × ratio of volumes = new pressure

$$785 \text{ torr} \times \frac{3.00 \text{ L}}{0.250 \text{ L}} = 9.42 \times 10^3 \text{ torr}$$

(3) 65.1 mL

The problem involves volumes and temperatures, which makes it a Charles' law problem. We know that V and T are directly related as long as T is expressed as a Kelvin or absolute temperature. We first have to convert °C to K. Any time you are asked to solve a gas law problem with a temperature involved, be sure to convert to K.

300°C + 273 = 573 K

100°C + 273 = 373 K

The temperature dropped from 573 K to 373 K. This means the volume must decrease also.

original volume × ratio of temperatures = new volume

$$100 \text{ mL} \times \frac{373 \text{ K}}{573 \text{ K}} = 65.1 \text{ mL}$$

(4) 334 mL

−50°C + 273 = 223 K

25°C + 273 = 298 K

The temperature increases so the volume must increase.

original volume × ratio of temperatures = new volume

$$250 \text{ mL} \times \frac{298 \text{ K}}{223 \text{ K}} = 334 \text{ mL}$$

(5) 31 mL
With a problem like this, it might be well to tabulate the data in an organized fashion before using the ratio technique.

	Initial Conditions	Final Conditions
P	0.75 atm	1.0 atm (standard)
V	45 mL	x mL
T	21°C	0°C (standard)
	294 K	273 K

The pressure increases so the volume will decrease. The temperature decreases, also causing the volume to decrease.

original volume × ratio of pressures × ratio of temperatures = new volume
(less than one)   (less than one)

$$45 \text{ mL} \times \frac{0.75 \text{ atm}}{1.0 \text{ atm}} \times \frac{273 \text{ K}}{294 \text{ K}} = 31 \text{ mL}$$

6. (1) 0°C, 273 K   (2) 1 atm, 760 torr   (3) 22.4 L   (4) same

7. (1) $1.34 \frac{g}{L}$

The formula wt of NO is $14.0 + 16.0 = 30.0 \frac{g}{mol}$

$$d = \frac{30.0 \text{ g}}{\text{mol}} \times \frac{1 \text{ mol}}{22.4 \text{ L}} = 1.34 \frac{g}{L}$$

(2) 82.6 L
Solve the equation for volume (L).

$$d = \frac{g}{L}$$

Multiply each side of the equation by "L".

$$(L)(d) = \frac{(g)(L)}{(L)}$$

Now divide each side of the equation by d.

$$\frac{(L)(d)}{(d)} = \frac{g}{d}$$

We now can substitute the values given for the density and the mass into the equation.

$$L = \frac{100 \text{ g} \times 1 \text{ L}}{1.21 \text{ g}} = 82.6 \text{ L HCN}$$

(3) 156 mL
Since the acetylene gas was collected over water, we have to subtract the partial pressure of water vapor at 21°C to find the partial pressure of acetylene alone. Then we can find the volume at STP.

$$P_{\text{total}} = P_{C_2H_2} + P_{H_2O}$$

750 torr = $P_{C_2H_2}$ + 18.6 torr (Appendix II)

$P_{C_2H_2}$ = 750 torr − 18.6 torr = 731.4 torr

Establish a table

	Initial Conditions	Final Conditions
P	731.4 torr	760 torr (standard)
V	175 mL	x mL
T	21°C 294 K	0°C (standard) 273 K

Using the ratio technique, we see that the pressure is increased and the temperature is reduced, both of which reduce the volume.

$$175 \text{ mL} \times \frac{731.4 \text{ torr}}{760 \text{ torr}} \times \frac{273 \text{ K}}{294 \text{ K}} = 156 \text{ mL}$$

8. 134 L

This is a volume-volume calculation and is an application of Avogadro's Law. Also, we need to consider whether there is a limiting reactant. First determine how many moles of NO$_{(g)}$ can be produced from each reactant.

$$4.00 \text{ mol } N_2 \times \frac{2 \text{ mol NO}}{1 \text{ mol } N_2} = 8.00 \text{ mol NO}$$

$$3.00 \text{ mol } O_2 \times \frac{2 \text{ mol NO}}{1 \text{ mol } O_2} = 6.00 \text{ mol NO}$$

Therefore O$_2$ is the limiting reactant and 6.00 mol NO will be produced in the reaction with 1.00 mol of N$_2$ left over.

The volume of 6.00 mol of NO at STP will be

$$6.00 \text{ mol} \times 22.4 \frac{\text{L}}{\text{mol}} = 134 \text{ L}$$

9. 52 L

The number of moles of starting substance is 7.0 moles NO$_2$.

Calculate the moles of NO, using the mole-ratio method.

$$7.0 \text{ mol NO}_2 \times \frac{1 \text{ mol NO}}{3 \text{ mol NO}_2} = 2.3 \text{ mol NO}$$

Convert moles of NO to L of NO. The moles of a gas at STP are converted to L by multiplying by the molar volume, 22.4 L per mole:

$$2.3 \text{ mol NO} \times \frac{22.4 \text{ L}}{\text{mol}} = 52 \text{ L NO}$$

10. 1.5×10^2 L (2 sig. figures)

   We must first calculate how many moles 15 g of H_2 represents and then solve the equation for "volume."

   $$15 \text{ g} \times \frac{1 \text{ mol } H_2}{2.0 \text{ g}} = 7.5 \text{ mol } H_2$$

   $$PV = nRT \quad \text{or} \quad V = \frac{nRT}{P}$$

   $$R = 0.0821 \text{ L-atm/mol K}$$

   $$V = \frac{7.5 \text{ mol} \times 0.0821 \frac{\text{L-atm}}{\text{mol K}} \times 298 \text{ K}}{1.2 \text{ atm}}$$

   $$= 1.5 \times 10^2 \text{ L}$$

11. Almond odor

   The molecular weight of arsine is 77.9 g/mol and that of cyanogen is 52 g/mol. Since the rate of effusion is inversely proportional to the square roots of their molecular weights, the hero will smell almonds first.

12. 185 moles O_2 and 5.92×10^3 g O_2

   We need to use the ideal gas equation to solve the problem. The data tabulated looks like this.

   $$PV = nRT$$

   $$P = 1500 \text{ psi} = (1500 \text{ psi}) \left(0.06805 \frac{\text{atm}}{\text{psi}}\right) = 102.1 \text{ atm}$$

   $$V = 43.8 \text{ L}$$

   $$R = 0.0821 \frac{\text{L-atm}}{\text{mol-K}}$$

   $$T = 72°F = 22°C = 295 \text{ K}$$

   $$n = \frac{PV}{RT} = \frac{(102.1 \text{ atm})(43.8 \text{ L})}{\left(0.0821 \frac{\text{L-atm}}{\text{mol-K}}\right)(295 \text{ K})} = 185 \text{ mol } O_2$$

   The mass of oxygen equals

   $$(185 \text{ mol } O_2)(32.0 \text{ g/mol}) = 5.92 \times 10^3 \text{ g}$$

13. 77.7 g/mole

   First we need to convert torr into atm and °C into K.

   $$740 \text{ torr} \times \frac{1 \text{ atm}}{760 \text{ torr}} = 0.974 \text{ atm}$$

   $$21°C + 273 = 294 \text{ K}$$

The ideal gas equation is: PV = nRT

Rearranging and using grams/mol wt = n, we have

$$\text{mol wt (M)} = \frac{g\ RT}{PV}$$

substituting into the equation:

$$M = \frac{3.48\ g \times 0.0821\ \cancel{L\text{-atm}} \times 294\ \cancel{K}}{0.974\ \cancel{atm} \times 1.11\ \cancel{L} \times \text{mol-}\cancel{K}}$$

$$= 77.7\ g/mole$$

This value varies somewhat from the accepted value of 77.9 g/mole because of rounding off several of the given values.

CHAPTER THIRTEEN
Water and the Properties of Liquids

SELF-EVALUATION SECTION

1. Many of the important characteristics of water are related to its physical properties. For example, the amount of heat required to change 1 gram of a solid into a liquid, called the (1) _____, is an unusually large amount for water, as is the amount of heat required to change 1 gram of liquid at its normal boiling point into a gas, called the (2) _____. The temperature at which ice begins to change into the liquid state is called the (3) _____ and the temperature at which the vapor pressure of water equals the atmospheric pressure of 1 atm is called the (4) _____.

Water reacts with numerous compounds to form useful products. For example, certain metallic oxides react with water to form bases and are known as (5) _____, while nonmetallic oxides form acids with water and are known as (6) _____. Calcium oxide (CaO) is a basic anhydride which is used in the cement industry, and sulfur trioxide (SO_3) is the acid anhydride of sulfuric acid (H_2SO_4). Other compounds contain water molecules as part of their crystalline structure and are called (7) _____. One often sees dramatic color changes when the water of hydration is removed from a hydrated salt. Cobalt chloride hexahydrate is rose-colored but turns sky-blue when heated. Some compounds are capable of absorbing water directly from the atmosphere. These substances are said to be (8) _____. Other substances, called (9) _____, absorb enough water to form a solution. Sodium hydroxide and P_2O_5 are examples of deliquescent substances. On the other end of the spectrum, there are a few crystalline materials that spontaneously lose water when exposed to the air. Such materials, such as Glauber's salt, are said to be (10) _____.

Water is not the only compound formed between hydrogen and oxygen. H_2O_2, which is named (11) _____, contains two oxygen atoms joined together by a single covalent bond. The charge on the peroxide ion is (12) _____, meaning that the oxidation number of each oxygen atom in the ion is (13) _____. The other electron from each oxygen atom is available to form (14) a covalent/an ionic bond to a hydrogen atom.

Peroxides are useful sources of (15) _____ gas. You can demonstrate this by observing a few ml of dilute H_2O_2 solution poured into the palm of your hand.

2. Water, which is the most common chemical substance around us, has been the subject of many experiments over the years. It is a simple molecule, H_2O, and yet its properties suggest that water is a very large molecule. We want to analyze how the structure of water influences such properties as boiling point and melting point. Examine Table 13.1 in the text prior to answering this section.

Fill in the blank space or circle the appropriate response.

A single water molecule consists of (1) _____ H atom(s) and (2) _____ O atom(s). The oxygen atom is the middle atom joined to each H atom by a(n) (3) ionic/covalent bond. The molecule is (4) straight end to end/bent in the middle with a bond angle of 105° between the two covalent bonds.

Oxygen is a very (5) electropositive/electronegative element and, as a result, the two covalent OH bonds are (6) nonpolar/polar. The bend in the molecule in conjunction with the polar covalent bonds makes water a (7) nonpolar/polar molecule. The oxygen atom carries a partial (8) _____ charge, and each hydrogen atom carries a partial (9) _____ charge.

Since each water molecule has the same unequal charge distribution, there will be an attraction between molecules. The oxygen side of one molecule will be attracted to the (10) _____ atom of another water molecule through a weak electrostatic bond. This type of attraction between an H atom and a highly electronegative atom is called a (11) _____ bond. The weak ionic association between water molecules produces the effect of water behaving as a large molecule. When we examine the physical properties of water, it is clear that water does not fit the expected pattern. The melting point and normal boiling point are (12) higher/lower than expected, as are the heat of fusion and heat of vaporization. Water acts as though it were a large bulky molecule rather than a small one with a molecular weight of 18. The bent structure and resulting polarity have a marked influence on the physical properties of water.

3. Identify each of the salt formulas as anhydrous or hydrates. Name each.
   (1) $CaSO_4$ _____
   (2) $CoCl_2 \cdot 6 H_2O$ _____
   (3) $NaC_2H_3O_2 \cdot 3 H_2O$ _____
   (4) K_2S _____
   (5) Na_3PO_4 _____

4. Identify the compounds listed as basic anhydrides or acidic anhydrides and write their reactions with water.
   (1) CaO (3) N_2O_5
   (2) SO_3 (4) Na_2O
   Write your answers below.

5. Write the formulas for the anhydrides of the following.
   (1) H_2SO_3, $HClO_4$, H_2CO_3
   (2) KOH, $Ba(OH)_2$, $Mg(OH)_2$
   Write your answers below.

6. Complete and balance the following reactions involving water as a reactant.
   (1) $K(s) + H_2O(l) \rightarrow H_2 \uparrow +$
   (2) $Al(s) + H_2O(steam) \rightarrow + Al_2O_3(s)$
   (3) $Fe(s) + H_2O(steam) \rightarrow H_2 \uparrow +$
   (4) $Cl_2 + H_2O(l) \rightarrow + HOCl$

7. Fill in the blank space.

   Natural fresh waters are usually not pure enough to drink and therefore must be treated. Removal of large objects is accomplished by (1) _____ whereas fine particles are removed by (2) _____ and (3) _____. The last step (4) _____ kills bacteria. If it contains dissolved magnesium and calcium salts, the water is said to be (5) _____. Three techniques used to soften hard water are (6) _____, (7) _____, and (8) _____. The process that uses zeolite to soften hard water is a type of (9) _____ _____ technique.

8. Fill in the blank space or circle the appropriate response.

   All substances in the liquid state are in the process of vaporizing. Some chemicals, such as acetone and ether, evaporate quickly, whereas others, such as mercury, do so slowly. The process of molecules going from the liquid state to the gas state is called (1) _____. In any sample of a liquid, (2) repulsive/attractive forces exist that must be overcome before a molecule can escape the liquid state. Even though the temperature of the liquid is uniform, not all of the molecules possess the same (3) kinetic/potential energy. Since the masses are constant, this suggests that the velocities of the molecules are (4) the same/different. Therefore, molecules at the surface of the liquid, which are moving faster than their neighbors, are able to overcome the attractive forces and escape to the gas state. After these molecules with greater kinetic energy leave the liquid state, the average kinetic energy of the remaining molecules in the liquid state is (5) lowered/raised. Therefore, the temperature (6) raises/drops, and we find that vaporization is a (7) warming/cooling process, which we know to be true from everyday experience. If the temperature drops as the average kinetic energy of the system goes down, you might wonder why the process goes on until the liquid is gone.

   In an open dish of water, for example, the water vaporizes completely. Where does the heat energy come from to maintain kinetic energy of the water so that vaporization proceeds? As you probably surmised, it comes from the surroundings.

   What will happen to our open dish of water if we place a large glass cover over it? The molecules of water at the surface of the liquid with sufficient kinetic energy will enter the gas state, and the liquid level in the dish will go down very slightly until an equilibrium situation is reached. In this equilibrium state, just as many water molecules leave the liquid state as enter it from the saturated vapor inside the glass cover. If we increase the temperature and, therefore, the kinetic energy, more liquid will vaporize until a new equilibrium is established. If the temperature is reduced, water molecules in the vapor or gas state will condense to the liquid state and the water level will increase. At any temperature, once equilibrium is established, as many molecules leave the liquid state as enter from the gas state. When an equilibrium state is reached, the vapor exerts a pressure just as any other gas does.

This pressure is known as (8) _____ and is an "internal pressure" or measure of the escape tendency of molecules from the liquid to the gas state. When the vapor pressure of a liquid reaches atmospheric pressure (760 torr), the liquid begins to (9) _____. For water, the temperature at which this happens is (10) _____. The vapor pressure of ethyl ether reaches 760 torr between 30°C and 40°C; and ethyl alcohol reaches 760 torr between 70°C and 80°C. It should be noted that atmospheric pressure is not always 760 torr, but varies with the weather and the elevation above sea level.

9. On the following heating curve, a solid substance at point A is heated until it reaches point E. Identify the various stages along the curve as requested.

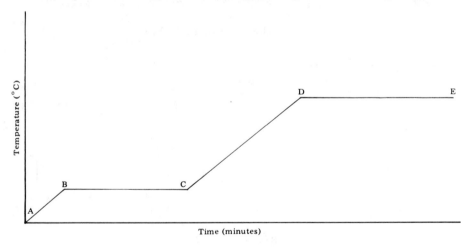

Identification of Process or Condition of State

(1) Point B _____

(2) Line BC _____

(3) Line CD _____

(4) Point D _____

(5) Line DE _____

Challenge Problems

10. An ice cube at 0°C weighs 8.75 g. Calculate how much heat energy (joules) is needed to melt the ice cube, raise the temperature of the resulting water to 100°C, and convert it to steam at 100°C.

    The heat of fusion is 335 J/g. The heat of vaporization is 2.26 kJ/g.

    Do your calculations here.

11. How many calories of heat energy are required to vaporize two 25.0 g pieces of solid ethanol, C_2H_5OH, at $-112°C$ given the following information.

    Ethanol, C_2H_5OH melting point $-112°C$
    boiling point $78.4°C$
    Heat of fusion 24.9 cal/g
    Heat of vaporization 204.3 cal/g
    specific heat, C_p, $2.49 \frac{\text{joule}}{\text{g}}$
    calories = (joules)$\left(0.239 \frac{\text{cal}}{\text{joule}}\right)$

    Do your calculations here.

RECAP SECTION

Chapter 13 is an enjoyable chapter to read. Even though there are many new terms, everything seems to fit together. We discussed a chemical compound that we are in contact with every day. Much of the discussion centered around the application of previously learned concepts. The importance of electronegativity in explaining the physical properties of water was good exercise for you. Since some of the topics in the chapter were not covered in the study guide, you may want to go back over the sections on the chemistry of water on your own. Now that you have learned some things about water by itself, let us begin to use water as we do in the laboratory to prepare chemical solutions. In the next unit, we discuss the principal use of water in chemistry—as a solvent for reagents.

ANSWERS TO QUESTIONS

1. (1) heat of fusion (2) heat of vaporization (3) melting point
   (4) normal boiling point (5) basic anhydrides (6) acid anhydrides
   (7) hydrates (8) hygroscopic (9) deliquescent
   (10) efflorescent (11) hydrogen peroxide (12) −2
   (13) −1 (14) a covalent (15) oxygen

2. (1) 2 (2) 1 (3) covalent
   (4) bent in the middle (5) electronegative (6) polar
   (7) polar (8) negative (9) positive
   (10) hydrogen (11) hydrogen (12) higher

3. (1) anhydrous Calcium sulfate (2) hydrate Cobalt(II) chloride hexahydrate (3) hydrate Sodium acetate trihydrate
   (4) anhydrous Potassium sulfide (5) anhydrous Sodium phosphate

4. (1) CaO—basic anhydride
   $CaO + H_2O \rightarrow Ca(OH)_2$

   (2) SO_3—acidic anhydride
   $SO_3 + H_2O \rightarrow H_2SO_4$

   (3) N_2O_5—acidic anhydride
   $N_2O_5 + H_2O \rightarrow 2HNO_3$

   (4) Na_2O—basic anhydride
   $Na_2O + H_2O \rightarrow 2NaOH$

5. (1) SO_2 Cl_2O_7 CO_2
   (2) K_2O BaO   MgO

6. (1) $2K(s) + 2H_2O(l) \rightarrow H_2 \uparrow + 2KOH(aq)$
   (2) $2Al(s) + 3H_2O(steam) \rightarrow 3H_2 \uparrow + Al_2O_3(s)$
   (3) $3Fe(s) + 4H_2O(steam) \rightarrow 4H_2 \uparrow + Fe_3O_4(s)$
   (4) $Cl_2 + H_2O(l) \rightarrow HCl(aq) + HOCl(aq)$

7. (1) screening (2) flocculation (3) sedimentation
   (4) disinfection (5) hard (6) precipitation
   (7) distillation (8) ion exchange (9) ion exchange

8. (1) vaporization (2) attractive (3) kinetic
   (4) different (5) lowered (6) drops
   (7) cooling (8) vapor pressure (9) boil
   (10) 100°C

9. (1) melting point
   (2) solid in equilibrium with liquid
   (3) all-liquid state, temperature begins to rise
   (4) boiling point
   (5) boiling liquid in equilibrium with gas

10. 26.4 kJ

    To melt the ice cube requires 335 J for each gram. Therefore,

    $$\left(335\frac{J}{g}\right) \times 8.75\, g = 2930 \text{ J}$$

    We now have water at 0°C and will need 4.184 J per gram of water per degree to raise the water temperature to boiling. Therefore,

    $$(8.75\, g)(100°C)\left(4.184\frac{J}{g°C}\right) = 3660 \text{ J}$$

    Notice how the units cancel to give us units of heat energy—joules. Now we must use 2.26 kJ for each gram of hot water to convert it into steam.

    $$(8.75\, g)(2.26 \text{ kJ/g}) = 19.8 \text{ kJ}$$

    The last step is to add up the individual values after converting to kJ.

    $$2.93 \text{ kJ} + 3.66 \text{ kJ} + 19.8 \text{ kJ} = 26.4 \text{ kJ}$$

11. 1.71×10^4 calories

    The first step is to melt the two 25.0 g pieces of solid ethanol at $-112°C$.

    $$(50.0 \text{ g})(24.9 \text{ cal/g}) = 1.25 \times 10^3 \text{ cal (3 sig. figures)}$$

    The temperature must be raised to the boiling point, 78.4°C. The specific heat is given in joules and must be changed to calories.

    (2.49 joule/g) (0.239 cal/joule) (50.0g) (temperature change)

    $$(2.49 \text{ joule/g})(0.239 \text{ cal/joule})(50.0 \text{ g})(190.4°) = 5.67 \times 10^3 \text{ cal}$$

    To vaporize the 50.0 g of liquid at the boiling point requires 204.3 cal/g

    $$(50.0 \text{ g})(204.3 \text{ cal/g}) = 1.02 \times 10^4 \text{ cal}$$

    Adding up the 3 values gives the final answer

    $$(1240 \text{ cal}) + (5670 \text{ cal}) + (10{,}200 \text{ cal}) = 1.71 \times 10^4 \text{ cal}$$

CHAPTER FOURTEEN
Solutions

SELF-EVALUATION SECTION

1. In the following statements, identify the (a) solution, (b) solute, and (c) solvent.

    (1) Household bleach is a 5% solution of sodium hypochlorite (NaClO) in water.

    a. _____ b. _____ c. _____

    (2) A 0.1 M iodine solution was prepared by dissolving crystals of iodine in carbon tetrachloride (CCl_4).

    a. _____ b. _____ c. _____

    (3) Air is composed primarily of two gases—oxygen and nitrogen. In round figures, air is 79% nitrogen and 21% oxygen.

    a. _____ b. _____ c. _____

    (4) Nickel coins in the United States are made from a nickel-copper alloy that is 75% copper and 25% nickel.

    a. _____ b. _____ c. _____

    (5) In order for certain species of fish to thrive in lakes and streams, the dissolved oxygen content of the water has to be 0.0005% or greater.

    a. _____ b. _____ c. _____

2. Solubility of salts in water.

    All nitrates are soluble in water.

    All chlorides, bromides, and iodides are soluble in water except those of silver, mercury(I), and lead(II).

    All carbonates and phosphates are insoluble in water except those of sodium, potassium, and ammonium.

    All sulfides are insoluble in water except those of ammonium sulfide and Group IA and Group IIA metals.

Place an s (for soluble) or i (for insoluble) next to each of these compounds.

(1) $CaBr_2$ _____   (11) $(NH_4)_2CO_3$ _____
(2) NH_4NO_3 _____   (12) BaI_2 _____
(3) Na_2S _____   (13) $LiI \cdot 3 H_2O$ _____
(4) $Mg_3(PO_4)_2$ _____   (14) $PbCl_2$ _____
(5) AgI _____   (15) KCl _____
(6) Fe_2S_3 _____   (16) Na_2CO_3 _____
(7) $ZrCl_4$ _____   (17) $Sn(NO_3)_4$ _____
(8) $NiCO_3$ _____   (18) Ag_2S _____
(9) $Al(NO_3)_3 \cdot 9 H_2O$ _____   (19) K_3PO_4 _____
(10) $HgBr$ _____   (20) $CaCO_3$ _____

3. Solubilities of salts in g/100 g H_2O.

$CaBr_2$	125 g at 0°C	$(NH_4)_2CO_3$	100 g at 15°C
NH_4NO_3	118 g at 0°C	$NaC_2H_3O_2$	119 g at 0°C
Na_2S	15.4 g at 10°C	$LiI \cdot 3 H_2O$	151 g at 0°C
$Na_2S_2O_3$	50 g at 20°C	KCl	34.7 g at 20°C
$MgSO_4$	26 g at 0°C	Na_2CO_3	7.1 g at 0°C
$Al(NO_3)_3$	63.7 g at 25°C	K_3PO_4	90 g at 20°C

Identify the solutions as saturated, unsaturated, or supersaturated. (All solutions are in 100 g H_2O.)

(1) 5 g of $CaBr_2$ at 0°C _____
(2) 100 g of NH_4NO_3 at 0°C _____
(3) 15.4 g of Na_2S at 10°C _____
(4) 55 g of $Na_2S_2O_3$ at 20°C _____
(5) 4 g of $MgSO_4$ at 0°C _____
(6) 51 g of $Al(NO_3)_3$ at 25°C _____
(7) 75 g of $(NH_4)_2CO_3$ at 15°C _____
(8) 125 g of $NaC_2H_3O_2$ at 0°C _____
(9) 38 g of $LiI \cdot 3 H_2O$ at 0°C _____
(10) 34.7 g of KCl at 20°C _____
(11) 6.5 g of Na_2CO_3 at 0°C _____
(12) 12 g of K_3PO_4 at 20°C _____

4. Fill in the blank space or circle the appropriate response.

Water molecules, which are very (1) polar/nonpolar, are (2) attracted/repulsed by other polar or ionic molecules or ions. Water molecules weaken the ionic forces that hold ions such as Na^+ and Cl^- together. Then the ions are pulled apart by the interaction with water as the water molecules surround the ions. The charged ions such as Na^+ and Cl^- then diffuse slowly away from the mass of undissolved salt as hydrated ions.

With most solid chemicals, an increase in temperature means an (3) _____ in solubility. However, a gaseous chemical always (4) _____ in solubility as the temperature increases. A temperature increase means that the kinetic energy of the gas molecules (5) _____ and their ability to associate with the solvent molecules (6) _____.

Pressure changes don't affect the solubility of solids greatly but gases show marked changes. The solubility of a gas is (7) inversely/directly proportional to the pressure of the gas above the liquid. Double the pressure means the solubility will (8) _____. Carbonated beverages are a good example.

5. List the four factors that influence the rate or speed at which a solid solute dissolves.

6. Before proceeding with exercises 6 through 10, be sure to study section 14.8 in the textbook.

   (1) How would you prepare 1500 g of a 2% sucrose (sugar) solution to be used for intravenous feeding in a hospital?

   Do your calculations here.

(2) How would you prepare 525 g of 0.1% salt solution (NaCl)?

Do your calculations here.

7. (1) Table wine is generally 12% alcohol by volume. How many mL of alcohol are contained in 250 mL of wine?

Do your calculations here.

(2) It is sometimes necessary in a chemistry laboratory to prepare a very corrosive solution of NaOH. How many grams of NaOH are contained in 375 g of 40% NaOH solution?

Do your calculations here.

8. (1) What is the molarity of a solution containing 4.2 moles of H_2SO_4 in 300 mL of solution?

Do your calculations here.

(2) What is the molarity of a KCl solution containing 1.7 moles of KCl in 3.0 L of solution?

Do your calculations here.

9. Calculate how many grams of chemical would be required to prepare the following solutions.

(1) 600 mL of 0.15 M NaF
(2) 4.00 L of 8.00 M NH_4NO_3
(3) 420 mL of 0.70 M $Ca(OH)_2$
(4) 250 mL of 0.94 M Na_2SO_4
(5) 0.50 L of 3.00 M $NaNO_3$

Atomic Weights

Na = 23.0	F = 19.0
S = 32.1	Ca = 40.1
O = 16.0	C = 12.0
N = 14.0	H = 1.0

Use the relationship

$$M = \frac{\text{g of solute}}{\text{formula wt of solute} \times \text{L of solution}}$$

or solve by dimensional analysis. Do your calculations here.

10. More experience in working with molarities of solutions.
   (1) What volume of 1.33 M $KMnO_4$ can be prepared from 180 g of $KMnO_4$?

$$\begin{array}{c} \underline{\text{Atomic Weights}} \\ K = 39.1 \\ Mn = 54.9 \\ O = 16.0 \end{array}$$

Do your calculations here.

(2) What volume of 0.10 M H_2SO_4 can be prepared from 147 g of H_2SO_4?

$$\text{Atomic Weights}$$
$$H = 1.0; S = 32.1; O = 16.0$$

Do your calculations here.

(3) Calculate the number of moles of hydrochloric acid in 4.0 L of 0.333 M HCl.

Do your calculations here.

(4) Calculate the number of moles contained in the following solutions:

250 mL of 0.22 M $K_2Cr_2O_7$
1500 mL of 1.4 M Na_2CO_3
3.15 L of 0.75 M $HClO_4$
857 mL of 0.66 M $CuSO_4 \cdot 5 H_2O$
50 mL of 0.25 M $Na_2S_2O_3$

Do your calculations in the space below.

(5) Using the relationship for dilutions, calculate how much (mL) concentrated reagent is necessary for the following solutions.

$$(volume_1)(M_1) = (volume_2)(M_2)$$

15 M NH_4OH to prepare 2.0 L of 3.0 M NH_4OH
14.6 M H_3PO_4 to prepare 300 mL of 0.15 M H_3PO_4
12 M HCl to prepare 6.0 L of 0.10 M HCl
18 M H_2SO_4 to prepare 500 mL of 0.45 M H_2SO_4

Do your calculations in the space below.

11. Water from the Great Salt Lake can have a salt concentration as high as 3.42 M (expressed as NaCl). High mountain spring water has a very low salt concentration, 0.00171 M. Potable, or drinkable, water has a maximum recommended level of 0.0171 M or 10 times that of spring water. What is the maximum volume of drinkable water that can be prepared from 500 mL of Great Salt Lake water using spring water for dilution? Assume the salt contribution from the spring water is negligible.

Do your calculations here.

12. Lead ion can be precipitated out of solution according to the following reaction.

$$Pb^{2+}_{(aq)} + Na_2CrO_4 \rightarrow PbCrO_4\downarrow + 2\,Na^+_{(aq)}$$

What weight of Na_2CrO_4 should be added to 10 L of solution that contains 2.78 g/L of Pb^{2+} ion?

Do your calculations here.

13. Colligative properties of solutions depend only on the (1) _____ of solute particles present. Therefore, 1 mole of sugar and 1 mole of alcohol, neither of which is ionic, will depress the freezing point or elevate the (2) _____ of a fixed amount of water by an (3) equal/unequal amount. These properties are usually expressed on the basis of a fixed amount of solvent, which is (4) _____, whether water or some other

solvent. Other colligative properties include (5) _____, in addition to freezing point depression and boiling point elevation. A mole of an ionic substance such as NaCl will lower the freezing point of a solution (6) _____ as much as a nonionic material since (7) _____ ions are produced for each mole of NaCl. A mole of $CaCl_2$ will lower the freezing point of water (8) _____ times as much as a mole of sugar or urea. For un-ionized and nonvolatile substances, molecular weights can be determined from either of two colligative properties, namely, (9) _____ and (10) _____.

The colligative properties of a solution can be considered in terms of vapor pressure. The vapor pressure of a pure liquid depends on the tendency of molecules to escape from its surface.

Thus, the presence of nonvolatile solute molecules at the surface will decrease the liquid's vapor pressure, since there are fewer solvent molecules at the surface when compared with a pure liquid. A liquid with a lower vapor pressure will consequently have a higher boiling point and a lower freezing point.

14. (1) How much ethylene glycol, $C_2H_4(OH)_2$, per kilogram of water is needed to lower the freezing from $0°C$ to $-20°C$?

The value for K_f is $1.86°C/mol/kg$. The equation is:

$$\Delta t_f = K_f \times \frac{\text{grams solute}}{\text{molecular wt solute}} \times \frac{1}{\text{kg solvent}}$$

Rearranging the equation to solve for grams of solute gives us

$$\text{g solute} = \frac{\Delta t_f \times \text{molecular wt} \times \text{kg solvent}}{K_f}$$

First determine the molecular wt of ethylene glycol and then make the necessary substitutions in order to solve the equation.

Do your calculations here.

(2) A sample of an organic compound weighing 1.50 g lowered the freezing point of 20.0 g of benzene by 2.75°C. The K_f of benzene is 5.1°C/mol/kg. Calculate the molecular or formula weight of the compound. The equation is:

$$\Delta t_f = K_f \times \frac{\text{grams solute}}{\text{molecular wt solute}} \times \frac{1}{\text{kg solvent}}$$

Rearranging the equation to solve for molecular weight gives us

$$\text{molecular wt} = \frac{K_f \times \text{g solute}}{\Delta t_f \times \text{kg solvent}}$$

Make the necessary substitutions and calculate the molecular wt of the unknown compound.

Do your calculations here.

15. Calculate the equivalent weights of the following acids and bases using the periodic table in the text.

   (1) $HClO_4$ _____
   (2) KOH _____
   (3) H_2SO_4 _____
   (4) $Ca(OH)_2$ _____
   (5) H_3PO_4 _____
   (6) H_2CO_3 _____
   (7) $Al(OH)_3$ _____
   (8) $Ba(OH)_2$ _____
   (9) HF _____
   (10) $NaOH$ _____
   (11) HCl _____
   (12) NH_4OH _____

16. Determine the Normality, N, of the following solutions whose concentrations are given as Molarity M.

   (1) 0.15 M H_3PO_4 _____
   (2) 0.1 M HCl _____
   (3) 0.45 M H_2SO_4 _____
   (4) 14.6 M H_3PO_4 _____
   (5) 18 M $HC_2H_3O_2$ _____
   (6) 0.6 M H_2CO_3 _____
   (7) 15 M NH_4OH _____
   (8) 0.7 M $Ca(OH)_2$ _____
   (9) 3.2 M $Al(OH)_3$ _____
   (10) 0.1 M KOH _____
   (11) 2.2 M $NaOH$ _____
   (12) 0.25 M $Ba(OH)_2$ _____

17. What volume of 0.125 N KOH is required to neutralize the following acid solutions?

   (1) 33 mL of 0.25 N H_2SO_4
   (2) 16.5 mL of 0.42 N HCl
   (3) 21.7 mL of 0.080 N H_2SO_4
   (4) 10.55 mL of 0.10 N HNO_3

Challenge Problems

18. Cadmium ion, which is a toxic metallic ion, can be precipitated from solution according to the following reaction.

$$Cd^{2+}_{(aq)} + K_3PO_{4(aq)} \rightarrow Cd_3(PO_4)_2 \downarrow + K^+$$

Balance the equation and, assuming no side reactions, determine what volume of stock 5.00 M K_3PO_4 solution should be diluted to 250 mL in order to precipitate 4.10 g Cd^{2+} ion that is contained in 600 mL of a waste solution.

Do your calculations here.

19. One gram (1.00 g) of glucose ($C_6H_{12}O_6$) dissolves in 1.10 mL of H_2O. What is the concentration in terms of weight percent and molality? Human blood normally contains approximately 0.090% glucose (weight percent). What is the molality of blood glucose?

Assume 1000 g of blood is equivalent to 1000 g of solvent.

Do your calculations here.

RECAP SECTION

After completing Chapter 14, you have learned many of the skills that a bench chemist or laboratory technician uses every day. These people often work with solutions; therefore, the knowledge of solution preparation and calculations involving solutions is important to them as well as to other scientists and technicians. The matter of proper preparation of solutions is critical in any work situation that deals with chemicals, such as nursing, agriculture, and food technology. There are many good review problems and questions at the end of the chapter in the text. Do any problems that are assigned and then try a few more. This is a good opportunity to sharpen up your problem-solving ability.

ANSWERS TO QUESTIONS AND SOLUTIONS TO PROBLEMS

1. (1) bleach, sodium hypochlorite, water
   (2) iodine solution, iodine, carbon tetrachloride
   (3) air, oxygen, nitrogen
   (4) coin alloy, nickel, copper
   (5) lake and river water, oxygen, water

2. (1) s   (2) s   (3) s   (4) i   (5) i   (6) i   (7) s
   (8) i   (9) s   (10) i   (11) s   (12) s   (13) s   (14) i
   (15) s   (16) s   (17) s   (18) i   (19) s   (20) i

3. (1) unsaturated   (2) unsaturated   (3) saturated   (4) supersaturated
   (5) unsaturated   (6) unsaturated   (7) unsaturated   (8) supersaturated
   (9) unsaturated   (10) saturated   (11) unsaturated   (12) unsaturated

4. (1) polar   (2) attracted   (3) increase   (4) decrease
   (5) increases   (6) decreases   (7) directly   (8) double

5. Four factors are particle size of solute, stirring, temperature, solution concentration.

6. (1) 30 g sucrose, 1470 g H_2O

   A 2% sucrose solution means that 2% of the total solution mass is sucrose.

   $$\frac{2}{100}(1500 \text{ g}) = 30 \text{ g sucrose}$$

   If 30 g of the total weight is sucrose, then the weight of water is

   1500 g − 30 g = 1470 g H_2O

   As a check, you can use the equation on page 334 of the text.

   $$\text{weight-percent} = \frac{\text{g solute}}{\text{g solute + g solvent}} \times 100\% = \frac{\text{g solute}}{\text{g solution}} \times 100\%$$

   $$= \frac{30 \text{ g}}{30 \text{ g} + 1470 \text{ g}} \times 100\% = 2.0\%$$

(2) 0.5 g NaCl, 524.5 g H_2O
A 0.1% solution means that 0.1% of 525 g is salt.

$$\left(\frac{0.1}{100}\right)(525 \text{ g}) = 0.525 \text{ g NaCl}$$

$$= 0.5 \text{ g (1 sig. figure)}$$

The mass of water is 525 g − 0.5 g = 524.5 g H_2O

7. (1) 30 mL of alcohol
You are asked to calculate a volume contained in a solution of a certain concentration; 12% of the total volume is alcohol.

$$\left(\frac{12}{100}\right)(250 \text{ mL}) = 30 \text{ mL alcohol}$$

As a check:

$$\frac{30}{250} \times 100\% = 12\% \text{ alcohol}$$

(2) 1.5×10^2 g of NaOH
Of the 375 g of solution, 40% is NaOH.

$$\left(\frac{40}{100}\right)(375 \text{ g}) = 1.5 \times 10^2 \text{ g of NaOH}$$

8. (1) 14 M H_2SO_4
For any problem involving molarity and quantities of chemicals needed to prepare for various solutions, it is wise to work from the definitions.

$$M = \frac{\text{mol of solute}}{\text{L of solution}}$$

From the problem, we have 4.2 moles of H_2SO_4 in 300 mL of solution.

$$M = \frac{4.2 \text{ mol}}{300 \text{ mL}} \times \frac{1000 \text{ mL}}{\text{L}} = 14 \text{ M}$$

(2) 0.57 M KCl

$$M = \frac{\text{mol of solute}}{\text{L of solution}} = \frac{1.7 \text{ mol}}{3.0 \text{ L}} = 0.57 \text{ M}$$

9. (1) 3.8 g NaF

formula wt of NaF = 23.0 + 19.0 = 42.0 g/mol

We need 600 mL of 0.15 M NaF.

$$M = \frac{g}{\text{formula wt} \times L}$$

Rearranging the equation to solve for g

g = M × formula wt × L

$$\text{g of NaF} = 0.15 \text{ M} \times 42.0 \frac{\text{g}}{\text{mol}} \times \frac{600 \text{ mL} \times 1 \text{ L}}{1000 \text{ mL}}$$

We should use the complete units for M so that our answer has the correct units of grams.

$$\text{g of NaF} = 0.15 \, \frac{\text{mol}}{\text{L}} \times 42.0 \, \frac{\text{g}}{\text{mol}} \times \frac{600 \, \text{mL} \times 1 \, \text{L}}{1000 \, \text{mL}}$$

$$= 3.8 \text{ g NaF}$$

(2) 2.56×10^3 g NH_4NO_3

formula wt of $NH_4NO_3 = 14.0 + (4 \times 1.0) + 14.0 + (3 \times 16.0)$

$$= 80.0 \, \frac{\text{g}}{\text{mol}}$$

You need 4.00 L of 8.00 M NH_4NO_3

g = M × formula wt × L

$$= 8.00 \, \frac{\text{mol}}{\text{L}} \times 80.0 \, \frac{\text{g}}{\text{mol}} \times 4.00 \, \text{L}$$

$$= 2.56 \times 10^3 \text{ g NH}_4\text{NO}_3$$

(3) 22 g $Ca(OH)_2$

formula wt of $Ca(OH)_2 = 40.1 + (2 \times 16.0) + (2 \times 1.0)$

$$= 74.1 \, \frac{\text{g}}{\text{mol}}$$

You are asked for 420 mL of 0.70 M $Ca(OH)_2$.

g = M × formula wt × L

$$= 0.70 \, \frac{\text{mol}}{\text{L}} \times 74.1 \, \frac{\text{g}}{\text{mol}} \times \frac{420 \, \text{mL} \times 1 \, \text{L}}{1000 \, \text{mL}}$$

$$= 22 \text{ g Ca(OH)}_2$$

(4) 33 g Na_2SO_4

formula wt of $Na_2SO_4 = (2 \times 23.0) + 32.1 + (4 \times 16.0)$

$$= 142.1 \, \frac{\text{g}}{\text{mol}}$$

You are asked for 250 mL of 0.94 M Na_2SO_4.

g = M × formula wt × L

$$= 0.94 \, \frac{\text{mol}}{\text{L}} \times 142.1 \, \frac{\text{g}}{\text{mol}} \times \frac{250 \, \text{mL} \times 1 \, \text{L}}{1000 \, \text{mL}}$$

$$= 33 \text{ g Na}_2\text{SO}_4$$

(5) 1.3×10^2 g $NaNO_3$

formula wt of $NaNO_3 = 23.0 + 14.0 + (3 \times 16.0)$

$$= 85.0 \, \frac{\text{g}}{\text{mol}}$$

You are asked for 0.50 L of 3.00 M NaNO$_3$.

$$g = M \times \text{formula wt} \times L$$

$$= 3.00 \, \frac{\text{mol}}{\cancel{L}} \times 85.0 \, \frac{g}{\cancel{\text{mol}}} \times 0.50 \, \cancel{L}$$

$$= 1.3 \times 10^2 \text{ g NaNO}_3$$

10. (1) 0.857 L or 857 mL

You are asked to solve the problem for a volume in units of L. With this in mind, write down the complete equation.

$$M = \frac{g}{\text{formula wt} \times L}$$

You now have to rearrange the equation so that "L" is in the numerator and isolated. Multiply both sides of the equation by "L."

$$(L)(M) = \frac{g \, \cancel{(L)}}{\text{formula wt} \times \cancel{L}}$$

Next, divide both sides by M

$$\frac{(L) \cancel{(M)}}{\cancel{(M)}} = \frac{g}{\text{formula wt} \times M}$$

$$L = \frac{g}{\text{formula wt} \times M}$$

Determine the formula wt of KMnO$_4$, substitute the data in the equation, and solve for "L."

$$39.1 + 54.9 + (4 \times 16) = 158 \, \frac{g}{\text{mol}}$$

$$L = \frac{180 \, \cancel{g} \times 1 \text{ L}}{158 \, \frac{\cancel{g}}{\cancel{\text{mol}}} \times 1.33 \, \cancel{\text{mol}}}$$

$$= 0.857 \text{ L or } 857 \text{ mL}$$

(2) 15 L H$_2$SO$_4$

$$H_2SO_4 = (2 \times 1.0) + 32.1 + (4 \times 16.0) = 98.1 \, \frac{g}{\text{mol}}$$

$$L = \frac{g}{\text{formula wt} \times M}$$

$$= \frac{147 \, \cancel{g} \times 1 \text{ L}}{98.1 \, \frac{\cancel{g}}{\cancel{\text{mol}}} \times 0.10 \, \cancel{\text{mol}}}$$

$$= 15 \text{ L}$$

(3) 1.3 moles of HCl

For this type of problem, go back to the simple first relationship since you are not asked a question involving grams or molecular weights. The defining equation for molarity is

$$M = \frac{\text{mol}}{L}$$

Therefore,

moles = L × M

This is what you have been asked to find in the above problem.

moles of HCl = 4.0 L × 0.333 $\frac{\text{mol}}{L}$ = 1.3 mol

(4) $K_2Cr_2O_7$ = 0.055 mole, Na_2CO_3 = 2.1 moles.

$HClO_4$ = 2.4 moles; $CuSO_4 \cdot 5 H_2O$ = 0.57 mole; $Na_2S_2O_3$ = 0.013 mole

$$\text{moles} = L \times M$$

$K_2Cr_2O_7$ moles = $\frac{250 \text{ mL} \times 1 \text{ L}}{1000 \text{ mL}}$ × 0.22 $\frac{\text{mol}}{L}$ = 0.055 mol

Na_2CO_3 moles = $\frac{1500 \text{ mL} \times 1 \text{ L}}{1000 \text{ mL}}$ × 1.4 $\frac{\text{mol}}{L}$ = 2.1 mol

$HClO_4$ moles = 3.15 L × 0.75 $\frac{\text{mol}}{L}$ = 2.4 mol

$CuSO_4 \cdot 5 H_2O$ moles = $\frac{857 \text{ mL} \times 1 \text{ L}}{1000 \text{ mL}}$ × 0.66 $\frac{\text{mol}}{L}$ = 0.57 mol

$Na_2S_2O_3$ moles = $\frac{50 \text{ mL} \times 1 \text{ L}}{1000 \text{ mL}}$ × 0.25 $\frac{\text{mol}}{L}$ = 0.013 mol

(5) NH_4OH = 400 mL, H_3PO_4 = 3.0 mL, HCl = 50 mL, H_2SO_4 = 13 mL

In each case, we must solve the relationship for the initial volume (V_1), which is the quantity of the concentrate reagent. Even though we dilute the reagent with water, the number of moles of chemical remains the same; the equation is a simple equality.

For the first problem, we dilute 15 M NH_4OH (ammonium hydroxide) to 2.0 L of 3.0 M NH_4OH.

Our equation is

$(V_1)(M_1) = (V_2)(M_2)$

$V_1 = \frac{(V_2)(M_2)}{M_1}$

Substituting in the equation

$$V_1 = \frac{(2.0 \text{ L}) (3.0 \text{ M})}{(15 \text{ M})}$$

V_1 = 0.4 L or 400 mL of 15 M NH₄OH diluted to 2.0 L with water

H_3PO_4 You have to change 300 mL into L before substituting in the equation.

$$V_1 = \frac{(0.300 \text{ L}) (0.15 \text{ M})}{(14.6 \text{ M})}$$

V_1 = 0.0031 L = 3.1 mL

HCl

$$V_1 = \frac{(6.0 \text{ L}) (0.10 \text{ M})}{(12 \text{ M})}$$

V_1 = 0.050 L = 50 mL

H_2SO_4 You have to change 500 mL into L first.

$$V_1 = \frac{(0.500 \text{ L}) (0.45 \text{ M})}{(18 \text{ M})}$$

V_1 = 0.013 L = 13 mL

11. 100 L
This is a dilution problem and we can use the equation

$$V_1 M_1 = V_2 M_2$$

where V_1 = 500 mL = 0.500 L
M_1 = 3.42 M = 3.42 mol/L
M_2 = 0.0171 M = 0.0171 mol/L
V_2 = Unknown final volume.

Substituting into the equation and solving for V_2

$$V_2 = \frac{(0.500 \text{ L}) (3.42 \text{ mol/L})}{0.0171 \text{ mol/L}}$$

= 100 L

12. 21.7 g Na_2CrO_4
The reaction equation is balanced as written and says that 1 mole of Pb^{2+} ion reacts with 1 mole of CrO_4^{2-} ion. So, we need to calculate how many moles of Pb^{2+} ion we have and from this calculate the weight of Na_2CrO_4 needed.

moles of Pb^{2+} = (wt. Pb^{2+}/atomic wt. Pb^{2+}) × 10 L

$$= \frac{(2.78 \text{ g/L})}{(207 \text{ g/mol})} \times 10 \text{ L} = 0.134 \text{ mol}$$

In order to precipitate 0.134 moles of Pb^{2+} ion, we must add an equal number of moles of Na_2CrO_4 which we can determine as follows:

formula weight of $Na_2CrO_4 = 46.0 + 52.0 + 64.0 = 162$ g/mol
wt. Na_2CrO_4 = (0.134 ~~mol~~) (162 g/~~mol~~) = 21.7 g

13. (1) number
    (2) boiling point
    (3) equal
    (4) 1 kg
    (5) vapor pressure lowering
    (6) about twice
    (7) 2 moles
    (8) about three
    (9) freezing point depression
    (10) boiling point elevation

14. (1) 6.7×10^2 g

    The problem states that the temperature differential is from 0°C to −20°C or a Δt of 20°. In addition, we have the value for K_f for water and the fact that we are dealing with 1 kilogram of solvent. We need to determine the molecular weight of ethylene glycol and then use the equation given.

    $$\begin{aligned}\text{Molecular wt of } C_2H_4(OH)_2 = 2 \times 12.0 &= 24.0 \\ 4 \times 1.0 &= 4.0 \\ 2 \times 16.0 &= 32.0 \\ 2 \times 1.0 &= 2.0 \\ &\,62.0 \text{ g/mole}\end{aligned}$$

    $$\text{g solute} = \frac{\Delta t_f \times \text{molecular wt} \times \text{kg solvent}}{K_f}$$

    Making the appropriate substitutions:

    $$= \frac{20°C \times 62.0 \text{ g/mol} \times 1 \text{ kg solvent}}{1.86°C/\text{mol}/\text{kg}}$$

    $= 6.7 \times 10^2$ g of ethylene glycol

    (2) 139 g/mol

    We are asked in this problem to determine the molecular weight of a compound from freezing point depression data. In order to solve the equation, we need to express the amount of solvent in units of kilograms and then make the necessary substitutions of the other values.

    20.0 g of benzene = 0.0200 kg
    $K_f = 5.1°C/\text{mol}/\text{kg}$
    $\Delta t_f = 2.75°C$
    grams of solute = 1.50 g

    $$\text{molecular wt} = \frac{K_f \times \text{g solute}}{\Delta t_f \times \text{kg solvent}}$$

    $$= \frac{5.1°C \times 1.50 \text{ g}}{\frac{\text{mol}}{\text{kg}} \times 2.75°C \times 0.0200 \text{ kg}}$$

    = 139 g/mol

15. (1) HClO₄ — 100.5 g/equiv. (7) Al(OH)₃ — 26.0 g/equiv.
    (2) KOH — 56.1 g/equiv. (8) Ba(OH)₂ — 85.7 g/equiv.
    (3) H₂SO₄ — 49.1 g/equiv. (9) HF — 20.0 g/equiv.
    (4) Ca(OH)₂ — 37.1 g/equiv. (10) NaOH — 40.0 g/equiv.
    (5) H₃PO₄ — 32.7 g/equiv. (11) HCl — 36.5 g/equiv.
    (6) H₂CO₃ — 31.0 g/equiv. (12) NH₄OH — 35.0 g/equiv.

16. (1) 0.45 N H_3PO_4 (7) 15 N NH_4OH
    (2) 0.1 N HCl (8) 1.4 N $Ca(OH)_2$
    (3) 0.90 N H_2SO_4 (9) 9.6 N $Al(OH)_3$
    (4) 43.8 N H_3PO_4 (10) 0.1 N KOH
    (5) 18 N $HC_2H_3O_2$ (11) 2.2 N NaOH
    (6) 1.2 N H_2CO_3 (12) 0.50 N $Ba(OH)_2$

17. (1) 66 ml of 0.125 N KOH
    (2) 55 ml
    (3) 14 ml
    (4) 8.4 ml

18. Balanced equation is $3Cd^{2+} + 2K_3PO_{4(aq)} \rightarrow Cd_3(PO_4)_2 \downarrow + 6K^+_{(aq)}$
    Volume of 5.00 M K_3PO_4 required is 4.86 mL. The equation states that 3 moles of Cd^{2+} require 2 moles of K_3PO_4 to precipitate. The number of moles of Cd^{2+} ion present is equal to

$$\frac{4.10 \text{ g} Cd^{2+}}{112.4 \text{ g/mol}} = 0.0365 \text{ mol } Cd^{2+}$$

We will need the following number of moles of K_3PO_4.

$$\text{Moles } K_3PO_4 = (0.0365 \text{ mol } Cd^{2+}) \frac{2 \text{ mol } K_3PO_4}{3 \text{ mol } Cd^{2+}} = 0.0243 \text{ mol}$$

We next use the dilution equation $V_1M_1 = V_2M_2$ where V_1 = unknown volume of stock 5.00 M solution and $M_1 = 5.00$ M.

Since 250 mL of the diluted K_3PO_4 is added to 600 mL of the Cd^{2+} solution, the final volume (V_2) will equal 850 mL (0.850 L).

$$V_2 = 250 \text{ mL} + 600 \text{ mL} = 850 \text{ mL} = 0.850 \text{ L}$$

$$M_2 = \frac{0.0243 \text{ mol}}{0.850 \text{ L}} = 0.0286 \text{ M}$$

$$V_1 = \frac{(0.850 \text{ L})(0.0286 \text{ M})}{5.00 \text{ M}}$$

$$= 0.00486 \text{ L} = 4.86 \text{ mL}$$

19. Concentrated solution is 5.05 M, blood is 0.50 m. The molecular weight of glucose is 180 g/mol. A solution of 1.00 g glucose in 1.10 mL of H_2O is equal to a percent concentration of

$$\frac{1.00 \text{ g glucose}}{(1.10 \text{ g}) + (1.00 \text{ g})} \times 100\% = 47.6\%$$

We can dissolve $\dfrac{1000}{1.10}$ g of glucose in 1 Kg of water.

This amount is equal to 909 g of glucose or $\dfrac{909 \text{ g}}{180 \text{ g/mol}}$ = 5.05 mol

The molality is therefore 5.05 m. Human blood has a concentration of 0.09%, or 90 g per 1000 g of blood. Using the assumption that 1000 g of blood is equivalent to 1000 g of solvent, the molality can be calculated as follows:

$$\frac{90 \text{ g}}{180 \text{ g/mol}} = 0.50 \text{ mol}$$

$$\text{molality} = \frac{0.50 \text{ mol}}{1 \text{ Kg solvent}} = 0.50 \; m$$

CROSSWORD PUZZLE 1

Across

2. A mixture without a uniform composition.
6. Yellow solid nonmetallic element (atomic no. 16).
7. An element that is ductile and malleable is a _____.
10. One shouldn't do crossword puzzles with a _____.
11. Binary compounds with a metallic element have names ending in _____.
12. Number of protons in hydrogen nucleus.
13. Chemical symbol for tantatum.
15. Chemical symbol for cobalt.
16. Elements whose d and f orbitals are being filled.
17. Formula for nitrogen oxide.
18. Light energy is sometimes considered to exist as a _____.
20. Chemical symbol for element number 95.
22. A pair of electrons forms a chemical _____ between two atoms.
24. Chemical symbol for osmium.
26. A positively charged ion.
27. Chemical symbol for element number 10.

Down

1. The answer to a mathematical problem is often called the _____.
3. Oxygen is an example of an _____, the basic building blocks of matter.
4. Nitrogen exists in what physical state as the uncombined element.
5. Chemical symbol for first element in period number 3.
8. Heat and light are forms of _____.
9. Element number 54.
10. Chemical symbol for element number 84.
13. The element whose chemical symbol is Sn.
14. The smallest indivisible unit of matter.
15. When elements combine chemically, they form a new substance called a _____.
19. A negatively charged ion is called an _____.
21. One Avogadro's Number of a substance dissolved in enough water to give 1 liter of solution is a 1 _____ solution.
23. Chemical symbol for element number 28.
25. Chemical symbol for element number 34.

Answers are found on page 236.

CHAPTER FIFTEEN
Ionization. Acids, Bases, Salts

SELF-EVALUATION SECTION

1. There are several different definitions of acids and bases according to various theories. Match the chemist's name with the appropriate phrase:
   - (1) Arrhenius
   - (2) Brønsted-Lowry
   - (3) Lewis
   - a. Acid is proton donor
   - b. Acid is electron pair acceptor
   - c. Acids are hydrogen-containing substances that dissociate to produce H^+

2. Identify the conjugate acid-base pairs in the following equations.
   - (1) $H_2BO_3^- + H_3O^+ \rightarrow H_3BO_3 + H_2O$
   - (2) $NH_3 + HBr \rightarrow NH_4^+ + Br^-$
   - (3) $HSO_4^- + H_3O^+ \rightarrow H_2SO_4 + H_2O$
   - (4) $HC_2H_3O_2 + H_2O \rightarrow H_3O^+ + C_2H_3O_2^-$
   - (5) $NH_4^+ + H_2O \rightarrow NH_3 + H_3O^+$

3. Complete and balance the following reactions between HCl and the chemicals given:
   - (1) HCl and NH_4OH
   - (2) HCl and Zn metal
   - (3) HCl and CaO
   - (4) HCl and Na_2CO_3

4. Complete the equation for the reaction of HCl and Al(OH)₃, an amphoteric hydroxide.
   $$HCl + Al(OH)_3 \rightarrow$$

5. Write balanced equations for the reaction between KOH and Zn metal and then for KOH and Al metal.
   Write your answer here.

6. Neutralization reactions. Identify the reactants as either *acids* or *bases*.

   (1) $NH_4OH + HClO_3 \rightarrow NH_4ClO_3 + H_2O$

   ———   ———

   (2) $Al(OH)_3 + H_3PO_4 \rightarrow AlPO_4 + 3\,H_2O$

   ———   ———

   (3) $H_2SO_4 + Ca(OH)_2 \rightarrow CaSO_4 + 2\,H_2O$

   ———   ———

   Dissociation reactions. Identify the acids and bases in each reaction according to the Brønsted-Lowry theory. Examine both the reactants and the products.

   (4) $NH_4^+ \;+\; H_2O \;\rightarrow\; NH_3 \;+\; H_3O^+$

   ———   ———   ———   ———

   (5) $C_2H_3O_2^- \;+\; H_2O \;\rightarrow\; HC_2H_3O_2 \;+\; OH^-$

   ———   ———   ———   ———

   (6) $Al^{3+} \;+\; H_2O \;\rightarrow\; Al(OH)^{2+} \;+\; H^+$

   ———   ———   ———   ———

7. In which of the following situations does the solution contain an electrolyte (E) or nonelectrolyte (NE)?

   (1) The solution conducts an electric current. ———
   (2) The solution contains only electrically neutral molecules. ———
   (3) The solution contains electrically charged ions. ———
   (4) The solution contains an acid, such as H_2SO_4. ———
   (5) The solution contains gaseous oxygen. ———
   (6) The solution is a nonconductor of electric current. ———
   (7) The solution contains molecules, such as acetic acid, which have reacted with water to produce ions. ———

8. Identify the following reactions as either dissociation (D) or ionization (I).

   (1) $NaCl \xrightarrow{H_2O} Na^+(aq) + Cl^-(aq)$ ———
   (2) $HC_2H_3O_2 + H_2O \rightarrow H_3O^+ + C_2H_3O_2^-$ ———
   (3) $NH_3(g) + H_2O \rightarrow NH_4^+ + OH^-$ ———
   (4) $Ca(OH)_2 \xrightarrow{H_2O} Ca^{2+}(aq) + 2\,OH^-(aq)$ ———
   (5) $HCl(aq) + H_2O \rightarrow H_3O^+ + Cl^-$ ———

9. Calculate the molarity of the ions in each of the salt solutions listed below. Consider each salt to be 100% dissociated.

   (1) 0.15 M $CaCl_2$ (4) 1.4 M LiI
   (2) 0.75 M $(NH_4)_2SO_4$ (5) 0.022 M $Al_2(SO_4)_3$
   (3) 3.3 M Na_3PO_4

10. Identify the following compounds as strong electrolytes (s) or weak electrolytes (w).

(1) HF
(2) $MgNO_3$
(3) NaOH
(4) $HC_2H_3O_2$
(5) $CaCl_2$
(6) HCl
(7) HClO
(8) H_3BO_3
(9) $HClO_4$
(10) NH_4OH

11. Rank these common solutions from strong acid to strong base. Indicate which solutions are acid, close to neutral, and basic using Table 15.4.

Solution	pH	Rank in order from acid to base
(1) 0.1 M HCl	1.0	
(2) Blood	7.4	
(3) Lemon juice	2.3	
(4) Drain cleaner	12.1	
(5) Vinegar	2.8	
(6) 0.1 M $HC_2H_3O_2$	2.9	
(7) Milk	6.6	
(8) Ammonia cleaner	8.2	
(9) Carbonated water	3.0	
(10) Tomato juice	4.1	

12. Let's do a little bit more with pH. The mathematical expression for pH indicates that we use the hydrogen ion concentration. For example, when $[H^+] = 1 \times 10^{-3}$ mol/L, the pH is 3.

When the $[H^+] = 1 \times 10^{-9}$ mol/L, the pH is 9.

As you can see, we derive the pH value from the exponent of 10 when the number in front is 1. When the number in front of the exponent is between 1 and 10, the pH is between the power of 10 given and the next lower power.

As an example:

$[H^+] = 6 \times 10^{-7}$ mol/L pH is between 7 and 6
$[H^+] = 5.4 \times 10^{-2}$ mol/L pH is between 2 and 1

(1) What is the pH of a solution whose $[H^+] = 1 \times 10^{-10}$ mol/L?
pH = _____

(2) What is the pH of a solution whose $[H^+] = 1 \times 10^{-7}$ mol/L?
pH = _____

(3) What is the pH of a solution whose $[H^+] = 4 \times 10^{-8}$ mol/L?
pH = _____

(4) What is the pH of a solution whose $[H^+] = 3.3 \times 10^{-5}$ mol/L?
pH = _____

13. A rather quick way to calculate a pH value for a solution of known H^+ ion concentration is to use a scale like the one in the text:

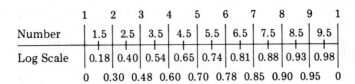

Let's use this scale to calculate the pH of a solution with $[H^+] = 6.5 \times 10^{-7}$.

$[H^+] = 6.5 \times 10^{-7}$

pH = this number − log of this number (must be between 1 and 10)

pH = 7 − log 6.5

From the log scale, log 6.5 = 0.81

pH = 7 − 0.81 = 6.19 (Answer)

Now you try the scale to find the pH of the following solutions:
(1) An acid fruit juice with a $[H^+] = 7 \times 10^{-5}$
(2) A detergent solution with a $[H^+] = 1.5 \times 10^{-9}$
(3) A hot breakfast drink with a $[H^+] = 2.5 \times 10^{-7}$

Write your answer here.

14. Indicate whether the following equations are *un-ionized* equations (U), *total ionic* equations (T) or *net ionic* equations (N).
   (1) $2\ OH^- + Mn^{2+} \rightarrow Mn(OH)_2 \downarrow$ _____
   (2) $2\ HCl(aq) + Ca(s) \rightarrow H_2 \uparrow + CaCl_2(aq)$ _____
   (3) $3\ NH_4OH(aq) + FeCl_3(aq) \rightarrow Fe(OH)_3 \downarrow + 3\ NH_4Cl(aq)$ _____
   (4) $2\ H^+(aq) + SO_4{}^{2-}(aq) + Mg \rightarrow$
   $\qquad H_2 \uparrow + Mg^{2+}(aq) + SO_4{}^{2-}(aq)$ _____
   (5) $2\ H^+(aq) + 2\ Cl^-(aq) + Na_2O(s) \rightarrow$
   $\qquad 2\ Na^+(aq) + 2\ Cl^-(aq) + H_2O$ _____
   (6) $2\ OH^-(aq) + Zn(s) \rightarrow ZnO_2{}^{2-}(aq) + H_2 \uparrow$ _____
   (7) $H^+ + OH^- \rightarrow H_2O$ _____
   (8) $Zn(OH)_2(s) + 2\ H^+(aq) + 2\ Cl^-(aq) \rightarrow$
   $\qquad Zn^{2+}(aq) + 2\ Cl^-(aq) + 2\ H_2O$ _____
   (9) $2\ H^+(aq) + CO_3{}^{2-}(aq) \rightarrow H_2O + CO_2 \uparrow$ _____
   (10) $H_2SO_4(aq) + MgCO_3 \rightarrow MgSO_4(aq) + H_2O + CO_2 \uparrow$ _____

15. Write balanced *un-ionized*, *total ionic*, and *net ionic* equations for the following two equations written in words.

   (1) Barium chloride solution plus silver nitrate solution react to form insoluble silver chloride plus barium nitrate solution.

   Un-ionized

   Total ionic

   Net ionic

   (2) Solid iron plus copper(II) sulfate solution react to form solid copper plus iron(II) sulfate solution.

   Un-ionized

   Total ionic

   Net ionic

16. 35 mL of 0.20 M HCl is required to titrate 50 mL of $Ca(OH)_2$ solution. What is the molarity of the base?

   Do your calculations here.

17. A 25.00 mL sample of $Ca(OH)_2$ solution required 21.35 mL of 0.3675 N H_2SO_4 for complete neutralization. What is the normality and the molarity of the calcium hydroxide? What is the molarity of the sulfuric acid?

   Do your calculations here.

Challenge Problems

18. Identify the conjugate acid-base pairs in the three step ionization of phosphoric acid, H_3PO_4. Write each equation and indicate the acid-base pairs.

   Do your work here.

19. How many grams, theoretically, of $BaSO_4$ will be precipitated when 35 mL of 0.15 M H_2SO_4 is added to 18 mL of 0.43 M $Ba(OH)_2$? What is the limiting reactant? If the actual yield is 1.0 g $BaSO_4$, what is the percentage yield?

   Do your calculations here.

20. You are an analyst working on product quality for a vinegar company. Your work this morning involves 10.00 mL samples of various batches of wine vinegar. The data table for the four titrations looks like this:

   Sample volume: 10.00 mL

	Batch A	B	C	D
Volume of 0.2500 N NaOH	33.05 mL	34.75 mL	32.80 mL	34.35 mL

   What is the average normality of acid in the vinegar? If the acidity is due to acetic acid ($HC_2H_3O_2$) what is the average molarity of the vinegar? Given that the density of the vinegar is 1.005 g/mL, what is the average weight/weight percentage concentration?

RECAP SECTION

Chapter 15 is an excellent chapter to be related to your laboratory experiments. You will be working with acids and bases, doing titrations, using the pH system, and examining the properties of electrolytes, nonelectrolytes, and colloids. Each of these topics is an important part of Chapter 15, and laboratory experiments will help clarify what was covered in lectures and homework problems. The last section on equation writing will be carried over to Chapter 17 when you examine oxidation-reduction reactions and try your hand at balancing all types of reaction equations.

ANSWERS TO QUESTIONS AND SOLUTIONS TO PROBLEMS

1. (1) c   (2) a   (3) b

2. (1) $H_2BO_3^-$ base, H_3BO_3 acid; H_3O^+ acid, H_2O base
   (2) NH_3 base, NH_4^+ acid; HBr acid, Br^- base
   (3) HSO_4^- base, H_2SO_4 acid; H_3O^+ acid, H_2O base
   (4) $HC_2H_3O_2$ acid, $C_2H_3O_2^-$ base, H_2O base, H_3O^+ acid
   (5) NH_4^+ acid, NH_3 base; H_2O base, H_3O^+ acid.

3. (1) $HCl_{(aq)} + NH_4OH_{(aq)} \rightarrow NH_4Cl_{(aq)} + H_2O$
   (2) $2HCl_{(aq)} + Zn_{(s)} \rightarrow ZnCl_{2\,(aq)} + H_2 \uparrow$
   (3) $2\,HCl_{(aq)} + CaO_{(s)} \rightarrow CaCl_{2\,(aq)} + H_2O$
   (4) $2\,HCl_{(aq)} + Na_2CO_{3\,(aq)} \rightarrow 2NaCl_{(aq)} + H_2O + CO_2 \uparrow$

4. $3HCl_{(aq)} + Al(OH)_{3\,(s)} \rightarrow AlCl_{3(aq)} + 3H_2O$

5. $2KOH + Zn_{(s)} \rightarrow Na_2ZnO_{2\,(aq)} + H_2 \uparrow$
   $6KOH + 2Al_{(s)} \rightarrow 2K_3AlO_{3\,(aq)} + 3H_2 \uparrow$

6. (1) NH_4OH = base $HClO_3$ = acid
   (2) $Al(OH)_3$ = base H_3PO_4 = acid
   (3) H_2SO_4 = acid $Ca(OH)_2$ = base
   (4) NH_4^+ = acid; H_2O = base; NH_3 = base; H_3O^+ = acid
   (5) $C_2H_3O_2^-$ = base; H_2O = acid; $HC_2H_3O_2$ = acid; OH^- = base
   (6) Al^{3+} = acid; H_2O = base; $Al(OH)^{2+}$ = base; H^+ = acid

7. (1) E   (2) NE   (3) E   (4) E   (5) NE   (6) NE   (7) E

8. (1) Dissociation (D). Ions are present in the crystalline salt, and the water molecules act only as the separation agent. Molten salt as well as a solution of salt conducts electricity.

   (2) Ionization (I). Water molecules are necessary to form ions in reaction with $HC_2H_3O_2$. Pure $HC_2H_3O_2$ is a very weak electrolyte.

   (3) Ionization (I). When ammonia gas dissolves in water, it undergoes reaction with water molecules to form a solution we call ammonium hydroxide (NH_4OH).

   (4) Dissociation (D).

   (5) Ionization (I). The bond in HCl(g) is predominantly covalent.

9. (1) 0.15 M Ca^{2+}, 0.30 M Cl^-
   (2) 1.5 M NH_4^+, 0.75 M SO_4^{2-}
   (3) 9.9 M Na^+, 3.3 M PO_4^{3-}
   (4) 1.4 M Li^+, 1.4 M I^-
   (5) 0.044 M Al^{3+}, 0.066 M SO_4^{2-}

10. (1) w (2) s (3) s (4) w (5) s
    (6) s (7) w (8) w (9) s (10) w

11. (1) (3) (5) (6) (9) (10) (7) (2) (8) (4)
    The first six (in order above) are acids, next two close to neutral, and the last two are basic.

12. (1) 10   (2) 7   (3) Between 8 and 7   (4) Between 5 and 4

13. (1) pH = 4.15
    $[H^+] = 7 \times 10^{-5}$
    pH = 5 − log 7

    From the log scale, log 7 = 0.85
    pH = 5 − 0.85 = 4.15

   (2) pH = 8.82
    $[H^+] = 1.5 \times 10^{-9}$
    pH = 9 − log 1.5

    From the log scale, log 1.5 = 0.18
    pH = 9 − 0.18 = 8.82

   (3) 6.6
    $[H^+] = 2.5 \times 10^{-7}$
    pH = 7 − log 2.5
    = 7 − 0.40 = 6.6

14. (1) N   (2) U   (3) U   (4) T   (5) T
    (6) N   (7) N   (8) T   (9) N   (10) U

15. (1) Un-ionized (U) $BaCl_2(aq) + 2\ AgNO_3(aq) \rightarrow 2\ AgCl\downarrow + Ba(NO_3)_2(aq)$
    Total ionic (T) $Ba^{2+}(aq) + 2\ Cl^-(aq) + 2\ Ag^+(aq) + 2\ NO_3^-(aq) \rightarrow$
    $2\ AgCl\downarrow + Ba^{2+}(aq)\ 2\ NO_3^-(aq)$
    Net ionic (N) $Ag^+(aq) + Cl^-(aq) \rightarrow AgCl\downarrow$

   (2) Un-ionized (U) $Fe(s) + CuSO_4(aq) \rightarrow Cu(s) + FeSO_4(aq)$
    Total ionic (T) $Fe(s) + Cu^{2+}(aq) + SO_4^{2-}(aq) \rightarrow$
    $Cu(s) + Fe^{2+}(aq) + SO_4^{2-}(aq)$
    Net ionic (N) $Fe(s) + Cu^{2+}(aq) \rightarrow Cu(s) + Fe^{2+}(aq)$

16. 0.070 M. Equation is $2\ HCl(aq) + Ca(OH)_2(aq) \rightarrow CaCl_2(aq) + 2\ H_2O$. The number of moles of acid used is twice the number of moles of base. Therefore, if we calculate the number of moles of acid used, we can use the mole-ratio method to determine the number of moles of $Ca(OH)_2$ in the 50 mL volume, and thus the concentration or molarity.

    number of moles of HCl = $0.035\ L \times \dfrac{0.20\ mol}{L}$ HCl
    = 0.0070 mol

    Using the mole ratio to determine the number of moles of $Ca(OH)_2$
    0.0070 mol HCl × $\dfrac{1\ mol\ Ca(OH)_2}{2\ mol\ HCl}$ = 0.0035 mol $Ca(OH)_2$

Therefore, 0.0035 mol of $Ca(OH)_2$ was present in 50 mL of $Ca(OH)_2$ solution.

Molarity of $Ca(OH)_2$

$$M = \frac{mol}{L} = \frac{0.0035 \text{ mol } Ca(OH)_2}{0.050 \text{ L}} = 0.070 \text{ M } Ca(OH)_2$$

17. The normality of the $Ca(OH)_2$ can be calculated from

$$V_A N_A = V_B N_B$$

substituting the data and solving for N_B

$$N_B = \frac{21.35 \text{ mL} \times 0.3675 \text{ N}}{25.00 \text{ mL}} = 0.3138 \text{ N } Ca(OH)_2$$

Since $Ca(OH)_2$ supplies 2 equivalents of OH per mole the conversion to molarity is

$$\frac{0.3138 \text{ equiv. } Ca(OH)_2}{L} \times \frac{1 \text{ mol } Ca(OH)_2}{2 \text{ equiv. } Ca(OH)_2} = 0.1569 \text{ mol/L}$$

The $Ca(OH)_2$ solution is 0.1569 M.

The molarity of the sulfuric acid can be found in a similar manner.

$$\frac{0.3675 \text{ equiv. } H_2SO_4}{L} \times \frac{1 \text{ mol } H_2SO_4}{2 \text{ equiv. } H_2SO_4} = 0.1838 \text{ mol/L}$$

The H_2SO_4 solution is 0.1838 M.

18. (1) $H_3PO_4 + H_2O \rightarrow H_3O^+ + H_2PO_4^-$

    (2) $H_2PO_4^- + H_2O \rightarrow H_3O^+ + HPO_4^{2-}$

    (3) $HPO_4^{2-} + H_2O \rightarrow H_3O^+ + PO_4^{3-}$

In each case H_3O^+ and H_2O are a conjugate acid-base pair. For reaction (1) H_3PO_4 is the acid and $H_2PO_4^-$ is the base. For reaction (2) $H_2PO_4^-$ is the acid and HPO_4^{2-} is the base while in (3) HPO_4^{2-} is the acid and PO_4^{3-} is the base.

19. H_2SO_4 is limiting, theoretical yield is 1.2 g $BaSO_4$, percentage yield is 83%.

Net ionic equation is $Ba^{2+} + SO_4^{2-} \rightarrow BaSO_4 \downarrow$

Number of moles $Ba^{2+} = \left(0.43 \frac{mol}{L}\right)(0.018 L) = 0.0077$

Number of moles $SO_4^{2-} = \left(0.15 \frac{mol}{L}\right)(0.035 L) = 0.0053$

Molecular weight $BaSO_4 = 233 \frac{g}{mol}$

Amount $BaSO_4 = \left(233 \frac{g}{mol}\right)(0.0053 \text{ mol}) = 1.2 \text{ g}$

Percentage yield $= \frac{\text{Actual yield}}{\text{Theoretical yield}} \times 100\% = \frac{1.0 \text{ g}}{1.2 \text{ g}} \times 100\% = 83\%$

20. The normality of each batch can be calculated from the titration data and the equation

$$N_A V_A + N_B V_B$$

$$N_A = \frac{N_B V_B}{V_A}$$

Substituting data for each batch into the equation we find that

	Batch A	B	C	D
Normality (N_A)	0.8263 N	0.8688 N	0.8200 N	0.8588 N

The average normality of the acid is: 0.8435 N

Since acetic acid ($HC_2H_3O_2$) has one equivalent of H^+ ion per mole, the average molarity will be the same value as the normality.

Molarity of $HC_2H_3O_2$ = 0.8435 M

The number of grams of acetic acid can be calculated from the molarity relationship.

$$0.8435 \; \frac{\text{moles}}{\text{L}} \times 60.06 \; \frac{\text{g}}{\text{mole}} = 50.66 \text{ g/L}$$

If the density of the vinegar is 1.005 g/mL then 1.0 L of vinegar will weigh

$$1.005 \text{ g/mL} \times 1000 \frac{\text{mL}}{\text{L}} = 1005 \text{ g}$$

The weight/weight percentage calculation is

$$\frac{50.66 \text{ g}}{1005 \text{ g}} \times 100\% = 5.041\% \; HC_2H_3O_2$$

CHAPTER SIXTEEN
Chemical Equilibrium

SELF-EVALUATION SECTION

Most of the examples and problems that we examine for Chapter 16 will have close counterparts in the text material. You should refer to the examples in the text as needed.

1. Write the following word equations for reversible systems in chemical equation form.
    (1) Two moles of sulfur dioxide gas plus one mole of oxygen gas react to produce two moles of sulfur trioxide gas.

    _____

    (2) Four moles of ammonia gas plus five moles of oxygen gas react to produce four moles of nitrogen oxide gas plus six moles of water vapor.

    _____

    (3) One mole of nitrogen gas plus three moles of hydrogen gas react to produce two moles of ammonia gas plus heat.

    _____

    (4) Three moles of oxygen gas plus heat react to produce two moles of ozone gas. One molecule of ozone consists of three atoms of oxygen.

    _____

    (5) One mole of carbon in the form of coke plus one mole of carbon dioxide gas plus heat react to produce two moles of carbon monoxide gas.

    _____

2. Using the Principle of Le Chatelier, predict the effect of changing various conditions on the following equilibrium systems.
    (1) $2\ SO_2\,(g) + O_2\,(g) \rightleftarrows 2\ SO_3\,(g)$
    Which direction, right or left, will the equilibrium be shifted if:

    a. the amount of SO_2 is decreased? _____

    b. the pressure is increased? _____

    c. a catalyst is added? _____

(2) $3 O_2(g) + \text{heat} \rightleftarrows 2 O_3(g)$
Which direction will the equilibrium be shifted if:

a. the amount of ozone is decreased? _____

b. the amount of ozygen is increased? _____

c. the reaction mixture is cooled? _____

d. the pressure is decreased? _____

(3) $C(s) + CO_2(g) + \text{heat} \rightleftarrows 2 CO(g)$
Which direction will the equilibrium be shifted if:

a. the amount of carbon is increased? _____

b. the reaction mixture is heated? _____

c. the amount of CO is increased? _____

d. the amount of CO_2 is increased? _____

e. the pressure is increased? _____

3. Write the equilibrium constant expression for the following reactions.

(1) $2 SO_2(g) + O_2(g) \rightleftarrows 2 SO_3(g)$

(2) $3 O_2(g) \xrightleftharpoons{\Delta} 2 O_3(g)$

(3) $C(s) + CO_2(g) \xrightleftharpoons{\Delta} 2 CO(g)$

(4) $HNO_2 \rightleftarrows H^+ + NO_2^-$

(5) $Ag^+ + 2 NH_3 \rightleftarrows Ag(NH_3)_2^+$

(6) $Al^{3+} + OH^- \rightleftarrows Al(OH)^{2+}$

(7) $PbSO_4(s) \rightleftarrows Pb^{2+} + SO_4^{2-}$

(8) $CH_4(g) + 2 O_2(g) \rightleftarrows CO_2(g) + 2 H_2O(g)$

(9) $4 HCl(g) + O_2(g) \rightleftarrows 2 H_2O(g) + 2 Cl_2(g)$

(10) $Cl_2(g) + H_2O \rightleftarrows HClO + HCl$

Write your answers here.

4. What is the value of K_{eq} for the reaction shown if a 1.0 L flask originally contains 0.5 moles of $SO_3(g)$ at 832° C and, at equilibrium, 50 percent of the SO_3 gas has decomposed?

$$2SO_3(g) \rightleftharpoons 2SO_2(g) + O_2(g)$$

5. (1) What is the concentration of hydrogen ions, H^+, in a 0.25 M solution of nitrous acid, HNO_2? $K_a = 4.5 \times 10^{-4}$

Do your calculations here.

(2) What is the concentration of hydrogen ions, H^+, in a 0.15 M solution of butyric acid, $CH_3(CH_2)_2COOH$? $K_a = 1.52 \times 10^{-5}$

Do your calculations here.

6. (1) What is the H^+ ion concentration and the OH^- ion concentration in a 0.1 M HBr solution? Assume the HBr is 100% ionized.

$K_w = 1 \times 10^{-14}$

Do your calculations here.

(2) What is the H^+ ion concentration and the OH^- ion concentration in a solution of pH 5?

Do your calculations here.

(3) What is the pH of a 1×10^{-4} M KOH solution? Assume that the KOH is 100% ionized.

Do your calculations here.

(4) Let's review in a short drill the relationships between pH, pOH, H^+, and OH^- concentrations.

a. The pH of a solution is 4.
   What is the pOH? _____
   What is the H^+ concentration? _____
   What is the OH^- concentration? _____

b. The pOH of a solution is 8.

   What is the OH⁻ concentration? _____

   What is the pH? _____

   What is the H⁺ concentration? _____

c. The H⁺ concentration is 1×10^{-11} mol/L.

   What is the pH? _____

   What is the pOH? _____

   What is the OH⁻ concentration? _____

(5) Fill in the blank space with a correct word or term.

A solution that consists of a weak acid or base and the corresponding salt, and which resists pH changes when diluted or when small amounts of strong acid or base are added is called a (a) _____ solution. In the acetic acid-sodium acetate buffer system, added H⁺ (from HCl) reacts with the (b) _____ ion to form un-ionized (c) _____. If hydroxide ion (OH⁻) from NaOH were added, it would react with the (d) _____ molecule to form two products (e) _____.

(6) Look at the compounds listed below. Match up the pairs that would be used for various buffer solutions.

   a. $HC_2H_3O_2$

   b. Na_2HPO_4

   c. H_2CO_3

   d. $H_2PO_4^-$

   e. $NaC_2H_3O_2$

   f. $NaHCO_3$

7. (1) Lead carbonate, $PbCO_3$, is a slightly soluble salt that dissociates according to the following reaction

$$PbCO_3(s) \rightleftarrows Pb^{2+} + CO_3^{2-} \qquad K_{sp} = 3.3 \times 10^{-14}$$

What is the concentration of CO_3^{2-} in a saturated lead carbonate solution?

Do your calculations here.

(2) Strontium sulfate, $SrSO_4$, is a slightly soluble salt that dissociates according to the following reaction:

$$SrSO_4(s) \rightleftarrows Sr^{2+} + SO_4^{2-} \qquad K_{sp} = 3.8 \times 10^{-7}$$

What is the concentration of Sr^{2+} in a saturated $SrSO_4$ solution?

Do your calculations here.

8. (1) Tin (II) carbonate, $SnCO_3$, has a solubility of 5.73×10^{-3} g/L. Calculate the K_{sp}.

Do your calculations here.

(2) One form of zinc sulfide, ZnS (sphalerite), has a solubility of 6.4×10^{-4} g/L. Calculate the K_{sp}.

Do your calculations here.

(3) Calcium fluoride, CaF_2, has a solubility of 1.6×10^{-3} g per 100 mL of H_2O. Calculate the K_{sp} remembering that CaF_2 produces two F^- ions and one Ca^{2+} ion when it dissociates. The dissociation equation is

$$CaF_2(s) \rightleftarrows Ca^{2+} + 2\ F^-$$

Do your calculations here.

9. Examine the formulas for the compounds shown. Make two lists ranking the acids from strongest to weakest and the salts from most soluble to least soluble. Then name all of the compounds.

HClO	$K_a = 3.5 \times 10^{-8}$
CaF_2	$K_{sp} = 3.9 \times 10^{-11}$
HCN	$K_a = 4.0 \times 10^{-10}$
$HC_2H_3O_2$	$K_a = 1.8 \times 10^{-5}$
$PbSO_4$	$K_{sp} = 1.3 \times 10^{-8}$
$Fe(OH)_3$	$K_{sp} = 6.0 \times 10^{-38}$

10. Calculate the H^+ ion concentration and the pH in a 0.1 M H_3PO_4 solution assuming the first ionization reaction is the primary reaction

$$H_3PO_4 \rightleftharpoons H^+ + H_2PO_4^- \quad K_a = 7.5 \times 10^{-3}$$

Also estimate the pH using the scale in either the text or the study guide.

Do your calculations here.

Challenge Problems

11. At 720 K, the K_{eq} for the following reaction has a value of 50.5.

$$H_{2(g)} + I_{2(g)} \rightleftharpoons 2HI_{(g)}$$

At the given temperature, what is the value for K_{eq} for the reverse reaction.

$$2HI_{(g)} \rightleftharpoons H_{2(g)} + I_{2(g)}$$

Calculate the equilibrium concentrations of all 3 species given a 1 L container and 0.0300 moles of HI gas at 720 K.

Do your calculations here.

12. Calculate the pH of a buffer solution prepared by adding 0.1 mole of sodium acetate to 500 mL of a solution labeled 0.4 M acetic acid. K_a is 1.8×10^{-5}.

   Do your calculations here.

13. By using net ionic hydrolysis reactions, show why solutions of the following salts are either acidic or basic.

   (1) NH_4NO_3 — acidic

   (2) K_2CO_3 — basic

   (3) NaCN — basic

   (4) $Al_2(SO_4)_3$ — acidic

RECAP SECTION

Chapter 16 discusses two of the most interesting topics in chemistry—chemical equilibrium and chemical kinetics. The material in the text is especially important for you to read and discuss with your instructor. The exercises in the study guide are primarily of a problem-solving nature; the concepts have been covered in the text. Once equilibrium has been reached in a chemical reaction system, you can use Le Chatelier's Principle to predict what will happen to the system if reaction conditions are altered. Such predictions have important practical value in the chemical industry. In addition, you have learned to use the equilibrium constant expression to evaluate chemical reactions for ion concentrations and equilibrium constants.

Scientists in industry and research draw on the concepts of Chapter 16 constantly to help them solve practical chemical problems.

ANSWERS TO QUESTIONS AND SOLUTIONS TO PROBLEMS

1. (1) $2\ SO_2(g) + O_2(g) \rightleftarrows 2\ SO_3(g)$

   (2) $4\ NH_3(g) + 5\ O_2(g) \rightleftarrows 4\ NO(g) + 6\ H_2O(g)$

   (3) $N_2(g) + 3\ H_2(g) \rightleftarrows 2\ NH_3(g) + heat$

   (4) $3\ O_2(g) + heat \rightleftarrows 2\ O_3(g)$

   (5) $C(s) + CO_2(g) + heat \rightleftarrows 2\ CO(g)$

2. (1) a. Decreasing the amount of SO_2 will produce a shift to the left. b. Increasing the pressure will cause a shift to the right, which is the side of fewer molecules. c. A catalyst will have no effect on the equilibrium.

(2) a. Decreasing the ozone concentration will shift the system to the right. b. So will increasing the amount of oxygen. c. Cooling the reaction will cause a shift to the left. d. Decreasing the pressure will cause a shift to the left.

(3) a. Increasing the amount of carbon will not shift the equilibrium. b. Heating the mixture will shift the equilibrium to the right. c. Increasing the amount of CO will shift the reaction to the left. d. Increasing the amount of CO_2 will produce the opposite effect. e. Increasing the pressure will drive the reaction to the left.

3. (1) $K_{eq} = \dfrac{[SO_3]^2}{[SO_2]^2[O_2]}$

(2) $K_{eq} = \dfrac{[O_3]^2}{[O_2]^3}$

(3) $K_{eq} = \dfrac{[CO]^2}{[CO_2]}$

Carbon is in the solid state; therefore, it does not change concentration. It can be left out of the equilibrium expression.

(4) $K_{eq} = \dfrac{[H^+][NO_2^-]}{[HNO_2]}$

(5) $K_{eq} = \dfrac{[Ag(NH_3)_2^+]}{[Ag^+][NH_3]^2}$

(6) $K_{eq} = \dfrac{[Al(OH)^{2+}]}{[Al^{3+}][OH^-]}$

(7) $K_{sp} = [Pb^{2+}][SO_4^{2-}]$

$PbSO_4$ is a solid; therefore, it is left out of the K_{sp} expression.

(8) $K_{eq} = \dfrac{[CO_2][H_2O]^2}{[CH_4][O_2]^2}$

(9) $K_{eq} = \dfrac{[H_2O]^2[Cl_2]^2}{[HCl]^4[O_2]}$

(10) $K_{eq} = \dfrac{[HClO][HCl]}{[Cl_2][H_2O]}$

4. $K_{eq} = 0.125$

The expression for the equilibrium constant, K_{eq}, would be:

$$K_{eq} = \dfrac{[SO_2]^2[O_2]}{[SO_3]^2}$$

If one-half of the SO_3 has decomposed at equilibrium, we can write our equilibrium conditions in tabular form.

Substance	Initial Conditions	Final Conditions
SO_3	0.5 M	0.25 M
SO_2	0 M	0.25 M
O_2	0 M	$\dfrac{0.25}{2}$ M

The equilibrium concentration of O_2 is one-half that of SO_2 according to the reaction. Substituting in to the expression for K_{eq} we have

$$K_{eq} = \frac{[SO_2]^2[O_2]}{[SO_3]^2} = \frac{[0.25]^2[0.125]}{[0.25]^2} = 0.125$$

5. (1) 1.05×10^{-2} M or mol/L

The formula for nitrous acid is HNO_2, so the first step is to write out the ionization equation. Then formulate the equilibrium expression.

$$HNO_2 \rightleftarrows H^+ + NO_2^-$$

$$K_a = \frac{[H^+][NO_2^-]}{[HNO_2]}$$

$$4.5 \times 10^{-4} = \frac{[H^+][NO_2^-]}{[HNO_2]}$$

Now you are ready to determine your strategy for making substitutions in the equation.

One way to keep everything straight is to set up a small table such as you did for gas law problems.

Substance	Initial Conditions	Final Conditions
H^+	0 M	X
NO_2^-	0 M	X
HNO_2	0.25 M	0.25 M $-$ X

Each molecule of HNO_2 that ionizes produces 1 H^+ ion and 1 NO_2^- ion. Therefore, let their equilibrium concentration be "X" so that the amount of HNO_2 left at equilibrium will be 0.25 M $-$ X. Substituting in the equation

$$4.5 \times 10^{-4} = \frac{[X][X]}{[0.25 - X]}$$

If X is small compared with 0.25 M, it is possible to simplify the equation.

$$4.5 \times 10^{-4} = \frac{X^2}{0.25}$$

$$1.1 \times 10^{-4} = X^2$$

To find the value for X, take the square root of 1.1 and of 10^{-4}. Refer to a math text or ask your instructor if you have difficulty with square roots.

Therefore,

$$X = 1.05 \times 10^{-2} \text{ M or mol/L} = [H^+]$$

(2) 1.0×10^{-2} M or mol/L

The formula for butyric acid is $CH_3(CH_2)_2COOH$, so the first step is to write out the ionization equation. After that formulate the equilibrium expression. The ionizable hydrogen in this case is shown on the right.

$$CH_3(CH_2)_2COOH \rightleftharpoons CH_3(CH_2)_2COO^- + H^+$$

$$K_a = \frac{[CH_3(CH_2)_2COO^-][H^+]}{CH_3(CH_2)_2COOH}$$

Set up a table as before.

Substance	Initial Conditions	Final Conditions
H^+	0 M	X
$CH_3(CH_2)_2COO^-$	0 M	X
$CH_3(CH_2)_2COOH$	0.15 M	0.15 − X

Substituting in our equilibrium expression, you have the following equation

$$1.52 \times 10^{-5} = \frac{[X][X]}{[0.15 - X]}$$

Assuming the X is small compared with 0.15, you can simplify your equation.

$$1.52 \times 10^{-5} = \frac{X^2}{0.15}$$

$$1.01 \times 10^{-4} = X^2$$

Taking the square root of both sides without difficulty is possible since the roots are divisible by 2. Therefore,

$$X = 1.0 \times 10^{-2} \text{ M or mol/L}$$

X is small compared to 0.15 so you were justified in making your assumption.

6. (1) $H^+ = 1 \times 10^{-1}$ M or mol/L, $OH^- = 1 \times 10^{-13}$ M or mol/L. The problem states that the HBr is completely ionized. The equation would be

$$HBr \rightarrow H^+ + Br^-$$

This means that the H^+ concentration is the same as the HBr solution concentration or 0.1 M. In scientific notation, 0.1 M would be 1×10^{-1} M. To determine the OH^- concentration, we need to use the ion product constant of water relationship, which states that

$$[H^+][OH^-] = K_w = 1 \times 10^{-14}$$

We know what the concentration of H^+ ion is, so we can make the following substitution into the ion product constant equation

$$[1 \times 10^{-1}][OH^-] = 1 \times 10^{-14}$$

Therefore:

$$[OH^-] = \frac{1 \times 10^{-14}}{1 \times 10^{-1}} = 1 \times 10^{-13} \text{ M or mol/L}$$

Remember, when we are dividing exponents we subtract the denominator exponent from the numerator exponent.

$$-14 - (-1) = -14 + 1 = -13$$

(2) $H^+ = 1 \times 10^{-5}$ M or mol/L, $OH^- = 1 \times 10^{-9}$ M or mol/L. We can determine the H^+ ion concentration directly from the pH and then use the ion product constant of water to calculate the OH^- ion concentration.

The pH value is a whole number, which means that the H^+ concentration is 1 times a negative amount of 10. The value of the exponent is the same as the pH value. Therefore, if the pH is 5, the H^+ ion concentration is 1×10^{-5} M or mol/L

Substituting this value into the ion product constant expression

$$[H^+][OH^-] = K_w = 1 \times 10^{-14}$$

$$[1 \times 10^{-5}][OH^-] = 1 \times 10^{-14}$$

$$[OH^-] = \frac{1 \times 10^{-14}}{1 \times 10^{-5}} = 1 \times 10^{-9} \text{ M or mol/L}$$

(3) pH = 10

There are at least two ways to solve this problem. By one method, we first need to find out what the H^+ concentration is. We can use the ion product constant expression to do this.

$$[H^+][OH^-] = 1 \times 10^{-14}$$

$$[H^+][1 \times 10^{-4}] = 1 \times 10^{-14}$$

$$[H^+] = \frac{1 \times 10^{-14}}{1 \times 10^{-4}} = 1 \times 10^{-10} \text{ M or mol/L}$$

We can easily convert this value to pH since the value in front of the exponential value is 1 (one). Therefore, the pH is 10, Another way to solve the problem is to use the relationship

$$pH + pOH = 14$$

Converting OH^- concentration into pOH, we have

$$1 \times 10^{-4} \text{ M} = pOH \text{ of } 4$$

Then we can substitute the pOH value

$$pH + 4 = 14$$
$$pH = 14 - 4 = 10$$

(4) a.   pOH = 10
$$[H^+] = 1 \times 10^{-4} \text{ M}$$
$$[OH^-] = 1 \times 10^{-10} \text{ M}$$

b. $[OH^-] = 1 \times 10^{-8}$ M
$$pH = 6$$
$$[H^+] = 1 \times 10^{-6} \text{ M}$$

c.   pH = 11
$$pOH = 3$$
$$[OH^-] = 1 \times 10^{-3} \text{ M}$$

(5) a. buffer   b. acetate   c. acetic acid
d. $HC_2H_3O_2$ e. $H_2O + C_2H_3O_2^-$

(6) a and e, b and d, c and f

7. (1) $CO_3^{2-} = 1.8 \times 10^{-7}$ M or mol/L
First, write the equilibrium expression with the appropriate value for the solubility product.

$$PbCO_3 \rightleftarrows Pb^{2+} + CO_3^{2-}$$

$$K_{sp} = 3.3 \times 10^{-14} = [Pb^{2+}][CO_3^{2-}]$$

Since $PbCO_3$ is a solid whose concentration is unchanging in the equilibrium system, you can cancel the term $[PbCO_3(s)]$ out of the equation. Since the concentration of Pb^{2+} is equal to the concentration of CO_3^{2-} at equilibrium, you can substitute "X" into the K_{sp} equation for both ion concentrations.

$$3.3 \times 10^{-14} = [X][X]$$
$$3.3 \times 10^{-14} = X^2$$
$$1.8 \times 10^{-7} = X$$

Therefore:
$$[Pb^{2+}] = [CO_3^{2-}] = 1.8 \times 10^{-7} \text{ M or mol/L}$$

(2) $Sr^{2+} = 6.2 \times 10^{-4}$ M or mol/L
The equation is
$$K_{sp} = 3.8 \times 10^{-7} = [Sr^{2+}][SO_4^{2-}]$$
Since
$$[Sr^{2+}] = [SO_4^{2-}] = X$$
Therefore:
$$3.8 \times 10^{-7} = [X][X]$$
$$3.8 \times 10^{-7} = X^2$$

Move the decimal point so that a square root can be taken.
$$38 \times 10^{-8} = X^2$$
$$6.2 \times 10^{-4} = X$$
Therefore:
$$[Sr^{2+}] = [SO_4^{2-}] = 6.2 \times 10^{-4} \text{ M or mol/L}$$

8. (1) $K_{sp} = 1.03 \times 10^{-9}$
First, calculate the formula weight of $SnCO_3$, using a list of atomic weights.
$$\text{formula wt of } SnCO_3 = (118.7) + (12.0) + (3 \times 16.0) = 178.7 \frac{g}{mol}$$

Next, determine the molarity of the saturated solution of $SnCO_3$, then substitute the values in the solubility product equation.
$$\text{Molarity} = \frac{g/L}{g/mol} = \frac{mol}{L}$$

Substituting in the above equation
$$M = \frac{5.73 \times 10^{-3} \text{ g/L}}{178.7 \text{ g/mol}} = 3.21 \times 10^{-5} \text{ M}$$
$$K_{sp} = [Sn^{2+}][CO_3^{2-}], [Sn^{2+}] = [CO_3^{2-}] = 3.21 \times 10^{-5} \text{ M}$$
$$= [3.21 \times 10^{-5}][3.21 \times 10^{-5}]$$
$$= 10.3 \times 10^{-10}$$
$$K_{sp} = 1.03 \times 10^{-9}$$

(2) $K_{sp} = 4.4 \times 10^{-11}$

$$\text{formula wt of } ZnS = 97.5 \frac{g}{mol}$$

$$M = \frac{g/L}{g/mol} = \frac{6.4 \times 10^{-4} \text{ g/L}}{97.5 \text{ g/mol}} = 6.6 \times 10^{-6} \text{ M}$$

$$K_{sp} = [Zn^{2+}][S^{2-}], [Zn^{2+}] = [S^{2-}] = 6.6 \times 10^{-6} \text{ M}$$
$$= [6.6 \times 10^{-6}][6.6 = 10^{-6}]$$
$$= 44 \times 10^{-12}$$
$$K_{sp} = 4.4 \times 10^{-11}$$

(3) $K_{sp} = 3.2 \times 10^{-11}$

First, calculate the formula weight of CaF_2, using a list of atomic weights.

$$\text{g-mol wt of } CaF_2 = 40.1 + (2 \times 19.0) = 78.1 \frac{g}{mol}$$

A solubility of 1.6×10^{-3} g per 100 mL of H_2O is the same as 1.6×10^{-2} g per L. Using this number, we can determine the molarity of a saturated solution of CaF_2.

$$\text{Molarity} = \frac{\text{moles}}{L}$$

Substituting in the above equation

$$M = \frac{1.6 \times 10^{-2} \text{ g/L}}{78.1 \text{ g/mol}} = 2.0 \times 10^{-4} \text{ M}$$

$$CaF_2(s) \rightleftharpoons Ca^{2+} + 2F^-$$

$$K_{sp} = [Ca^{2+}][F^-]^2$$
$$= [2.0 \times 10^{-4}][2 \times (2.0 \times 10^{-4})]^2$$

Remember there are two F^- ions for each Ca^{2+} ion.

$$K_{sp} = [2.0 \times 10^{-4}][4.0 \times 10^{-4}]^2$$
$$= 32 \times 10^{-12}$$
$$= 3.2 \times 10^{-11}$$

9. Acids $HC_2H_3O_2$, acetic acid, is strongest
           HClO, hypochlorous acid, is next
           HCN, hydrocyanic, is weakest

    Salts CaF_2, calcium fluoride, is most soluble
           $PbSO_4$, lead(II) sulfate, is next
           $Fe(OH)_3$, iron(III) hydroxide, is least soluble

10. H^+ concentration equals 2.74×10^{-2} M, pH is approximately 1.56. Establish a table as before.

Substance	Initial Conditions	Final Conditions
H_3PO_4	0.1 M	0.1 M − X
$H_2PO_4^-$	0 M	X
H^+	0 M	X

Substitute the final condition values into the expression for K_a and assume X has a small value compared to the initial concentration of H_3PO_4

$$K_a = 7.5 \times 10^{-3} = \frac{[H^+][H_2PO_4^-]}{[H_3PO_4]} = \frac{[X][X]}{[0.1-X]}$$

$$7.5 \times 10^{-3} = \frac{X^2}{0.1}$$

$$[H^+] = X = \sqrt{7.5 \times 10^{-4}} = 2.74 \times 10^{-2}$$

$$pH = -\log[H^+] = -(+0.44 - 2.00) = 1.56$$

11. $K_{eq} = 1.98 \times 10^{-2}$, $[H_2] = [I_2] = 4.22 \times 10^{-3}$ M, $[HI] = 2.16 \times 10^{-2}$ M. Quadratic formula yields $[H_2] = [I_2] = 5.85 \times 10^{-3}$ M, $[HI] = 1.83 \times 10^{-2}$ M. The k_{eq} for the reverse reaction is the reciprocal of the K_{eq} for the forward reaction.

$$K_{eq}(\text{reverse}) = \frac{1}{K_{eq}(\text{forward})} = \frac{1}{50.5} = 1.98 \times 10^{-2}$$

The initial concentration of HI is 0.0300 M. Set up a table of values as before

Substance	Initial Conditions	Final Conditions
HI	0.0300 M	$0.0300 - X$
H_2	0	$\frac{X}{2}$
I_2	0	$\frac{X}{2}$

One mole of HI decomposes to produce 1/2 mole each of H_2 and I_2.

$$K_{eq} = \frac{[H_2][I_2]}{[HI]^2} = \frac{\left[\frac{X}{2}\right]\left[\frac{X}{2}\right]}{[0.0300-X]^2} = \frac{\left[\frac{X^2}{4}\right]}{0.0009} = 1.98 \times 10^{-2}$$

Assume X is small compared to 0.0300 M.

Simplifying

$$(1.98 \times 10^{-2})(9 \times 10^{-4})(4) = X^2$$

$$X^2 = 7.13 \times 10^{-5}$$

$$X = 8.44 \times 10^{-3}$$

If $X = 8.44 \times 10^{-3}$, then the concentrations of H_2 and I_2 are X/2 or 4.22×10^{-3} M. The concentration of HI is 0.0300 M $-$ 0.00844 M or 0.0216 M.

The value for X is not really small compared to 0.0300 M. Using the quadratic formula, a value for X of 1.17×10^{-2} is obtained rather than 8.44×10^{-3}. In this case the concentrations of H_2 and I_2 would be calculated to be 5.85×10^{-3} M and the concentration of HI would be 0.0183 M.

12. pH = 4.4
The equation is $HC_2H_3O_2 \rightleftharpoons H^+ + C_2H_3O_2$

$$K_a = \frac{[H^+][C_2H_3O_2^-]}{[HC_2H_3O_2]} = 1.8 \times 10^{-5}$$

0.1 mole sodium acetate furnishes 0.1 mole acetate ion.

$$NaC_2H_3O_2 \rightarrow Na^+ + C_2H_3O_2^-$$

$$\frac{0.1 \text{ mol}}{500 \text{ mL}} = \frac{0.2 \text{ mol}}{1000 \text{ mL}} = \frac{0.2 \text{ mol}}{L}$$

Substituting into the equation we have

$$\frac{[H^+][0.2]}{[0.4]} = 1.8 \times 10^{-5}$$

$$H^+ = \frac{(1.8 \times 10^{-5})(0.4)}{0.2} = 3.6 \times 10^{-5} M$$

$$pH = -\log[H^+] = -(0.56 - 5) = 4.4$$

The pH can also be calculated from an equation named the Henderson-Hasselbach equation

$$pH = pKa + \log\frac{[salt]}{[acid]}$$

The answer will be the same either way.

13. (1) $NH_4^+ + H_2O \rightleftharpoons NH_4OH + H^+$
    (2) $CO_3^{2-} + 2H_2O \rightleftharpoons H_2CO_3 + 2OH^-$
    (3) $CN^- + H_2O \rightleftharpoons HCN + OH^-$
    (4) $2Al^{3+} + 6H_2O \rightleftharpoons 2Al(OH)_3 + 6H^+$

WORD SEARCH 2

In the given matrix, find the terms that match the following definitions. Terms may be horizontal, vertical, or on the diagonal. They may also be written forward or backward. Answers are found on page 237.

1. A principle that states what happens to an equilibrium system when the conditions are altered.
2. A solution that resists changes in pH.
3. An experimental technique for measuring the volume of one reagent required to react with a measured amount of another reagent.
4. The formation of ions.
5. A dynamic state where two or more opposing processes are taking place at the same time and at the same rate.
6. A homogeneous mixture of two or more substances.
7. A substance whose aqueous solution conducts electricity.
8. Capable of mixing and forming a solution.
9. The H_3O^+ ion.
10. A solution containing 1 mole of solute per liter of solution is 1.0 _____.
11. A solution containing dissolved solute in equilibrium with undissolved solute.
12. Behaving chemically as either an acid or base.
13. The substance present to the largest extent in a solution.
14. The number of equivalent weights of solute per liter of solution.
15. Incapable of mixing.
16. The substance that is dissolved in a solvent to form a solution.
17. Solution containing a relatively small amount of solute.

A	E	G	Y	T	F	E	H	Y	K	B	D	I	S	P	D	N	J
G	P	I	S	O	L	U	T	E	O	Q	F	Y	D	I	L	K	Z
X	T	D	N	O	A	I	S	Y	C	I	E	V	L	A	H	P	O
C	K	E	B	G	L	O	M	R	L	X	R	U	N	F	O	J	Z
U	M	T	V	A	E	V	Z	M	A	O	T	E	Q	B	Q	Y	M
S	F	A	M	H	C	R	E	D	I	E	R	E	F	F	U	B	Q
D	N	R	U	S	H	E	R	N	C	S	Q	T	U	W	M	J	P
C	O	U	I	J	A	L	B	P	T	L	C	R	C	G	H	Z	W
N	I	T	N	U	T	B	I	N	E	H	A	I	V	E	E	Y	I
X	T	A	O	W	E	I	C	T	O	L	R	T	B	A	L	O	Q
K	A	S	R	D	L	C	V	M	O	E	W	M	Z	L	V	E	N
V	R	G	D	L	I	S	E	M	T	B	L	N	E	C	E	B	L
O	T	F	Y	X	E	I	I	O	N	I	Z	A	T	I	O	N	R
Y	I	G	H	W	R	M	H	Z	A	X	U	F	A	K	T	T	F
I	T	S	V	A	X	P	U	N	O	I	T	U	L	O	S	G	J
J	P	L	Q	Z	M	U	I	R	B	I	L	I	U	Q	E	E	T
M	C	X	S	A	H	T	A	U	D	F	N	Y	P	R	B	R	W

CHAPTER SEVENTEEN
Oxidation-Reduction

SELF-EVALUATION SECTION

1. In the following reactions, identify the element that is *reduced*, the element that is *oxidized*, the *reducing agent*, and the *oxidizing agent*.

   Oxidation Information

   Elements in the free state have an oxidation number of 0.
   Hydrogen is usually +1.
   Oxygen is usually −2.
   Metallic elements in ionic compounds have a positive charge.
   Group IA metals are always +1.

   (1) $Cr_2O_3 + 3\ H_2(g) \xrightarrow{\Delta} 2\ Cr + 3\ H_2O$

   (2) $2\ H_2(g) + O_2(g) \xrightarrow{\Delta} 2\ H_2O$

   (3) $2\ HNO_2(aq) + 2\ HI(aq) \rightarrow I_2(s) + 2\ NO \uparrow + 2\ H_2O$

   (4) $2\ Na(s) + H_2O \rightarrow 2\ NaOH(aq) + H_2 \uparrow$

   (5) $5\ NaBr + NaBrO_3 + 3\ H_2SO_4 \rightarrow 3\ Br_2 + 3\ Na_2SO_4 + 3\ H_2O$

   Write your answers here.

2. Using the oxidation number information given for question 1, balance the following un-ionized and ionic equations using the electron gain-and-loss technique. Place correct coefficients in front of each formula.

   (1) $HNO_2(aq) + HI(aq) \rightarrow NO(g) + I_2(s) + H_2O$
   (2) $Mn^{4+} + Cl^-(aq) \rightarrow Cl_2\uparrow + Mn^{2+}(aq)$
   (3) $ClO_3^-(aq) + SnO_2^{2-}(aq) \rightarrow Cl^-(aq) + SnO_3^{2-}(aq)$
   (4) $NH_3(g) + O_2(g) \rightarrow NO(g) + H_2O(g)$
   (5) $H^+(aq) + Br^-(aq) + SO_4^{2-}(aq) \rightarrow Br_2 + SO_2\uparrow + H_2O$

   Write your answers here.

3. Balance the following redox equations by the ion-electron method. For acidic conditions use H^+ and H_2O to balance H and O. For basic conditions use OH^- and H_2O.

   (1) $Zn(s) + NO_3^- \rightarrow Zn^{2+} + NH_4^+$ (acidic)
   (2) $ClO^- + I^- \rightarrow IO_3^- + Cl^-$ (basic)

   Write your answers here.

4. Using the partial list of the electromotive series of metals, answer the following questions. For any reaction that proceeds write a balanced equation.

$Mg \rightarrow Mg^{2+} + 2e^-$

$Al \rightarrow Al^{3+} + 3e^-$

$Zn \rightarrow Zn^{2+} + 2e^-$

$Fe \rightarrow Fe^{2+} + 2e^-$

$Ni \rightarrow Ni^{2+} + 2e^-$

$Sn \rightarrow Sn^{2+} + 2e^-$

$Pb \rightarrow Pb^{2+} + 2e^-$

$H_2 \rightarrow 2 H^+ + 2e^-$

$Cu \rightarrow Cu^{2+} + 2e^-$

$Ag \rightarrow Ag^+ + e^-$

(1) Will a chemical reaction occur when a $Pb(NO_3)_2$ solution (Pb^{2+} + 2 NO_3^-) is placed in a *copper* container?

(2) Will a chemical reaction occur when a $ZnCl_2$ solution (Zn^{2+} + 2 Cl^-) is placed in a *magnesium* container?

(3) Will a chemical reaction occur when a HCl solution (H^+ + Cl^-) is placed in a *tin* can?

(4) Will a chemical reaction occur when an $AgNO_3$ solution (Ag^+ + NO_3^-) is placed in an *aluminum* container?

Do your equations here.

5. Aluminum metal is produced almost entirely by the electrolysis of Al_2O_3. The simplified reactions are

$$Al^{3+} + 3e^- \rightarrow Al$$
$$O^{2-} + C \rightarrow CO \uparrow + 2e^-$$

In the sketch below, indicate the direction of ion migration and balance the overall equation.

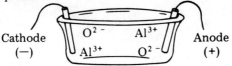

Write your answers here.

6. (1) Write the anode and cathode reactions for the electrolysis of molten $NiBr_2$ with inert electrodes. Then write the balanced overall reaction.

Write your answers here.

(2) Write anode, cathode, and the overall reaction for the electrolysis of molten LiCl with inert electrodes.

Write your answers here.

7. Identify which half-reaction of each pair occurs at the anode and which half-reaction occurs at the cathode.

   (1) $6 \text{ Fe}^{2+} \rightarrow 6 \text{ Fe}^{3+} + 6e^-$
   $Cr_2O_7^{2-} + 14 \text{ H}^+ + 6e^- \rightarrow 2 \text{ Cr}^{3+} + 7 \text{ H}_2\text{O}$

   (2) $2 \text{ H}^+ + \text{H}_2\text{O}_2 + 2e^- \rightarrow 2 \text{ H}_2\text{O}$
   $2 \text{ I}^- \rightarrow \text{I}_2 + 2e^-$

   (3) $\text{Fe}^0 \rightarrow \text{Fe}^{2+} + 2e^-$
   $Ni_2O_3 + 3 \text{ H}_2\text{O} + 2e^- \rightarrow 2 \text{ Ni(OH)}_2 + 2 \text{ OH}^-$

   Write your answers here.

8. Fill in the blank space with a correct term.

   The type of cell that uses chemical reactions to produce electrical energy is called a (1) _____ cell. The other type of cell, (2) _____, uses electrical energy to produce a chemical change. A dry cell flashlight battery is an example of a(n) (3) _____.

9. The following reactions are involved in the operation of the lead storage battery. Which is the oxidation reaction and which is the reduction reaction? Give the overall balanced cell reaction.

   (1) $\text{Pb} \longrightarrow \text{Pb}^{2+} + 2e^-$
   (2) $\text{PbO}_2 + 4\text{H}^+ + 2e^- \longrightarrow \text{Pb}^{2+} + 2\text{H}_2\text{O}$

Challenge Problems

10. The reaction used to measure the biochemical oxygen demand of a water sample involves a 0.025N solution of sodium thiosulfate according to the following reaction.

$$2Na_2S_2O_3 \cdot 5H_2O + I_2 \rightarrow Na_2S_4O_6 + 2NaI + 10H_2O$$

How much sodium thiosulfate pentahydrate should be weighed out to prepare 1 L of a 0.025N solution for the above reaction?

Do your calculations here.

11. Balance the following oxidation-reduction equations. The reactions conditions are either acidic or basic, and H^+ and OH^- plus H_2O should be used accordingly to balance the reactions.

   (1) $MnO_4^- + VO^{2+} \rightarrow VO_2^+ + Mn^{2+}$ (acid)

   (2) $P_4 \rightarrow PH_3 + H_2PO_2^-$ (basic solution)

   (3) $MnO_2 + SO_3^{2-} \rightarrow SO_4^{2-} + Mn(OH)_2$ (basic solution)

   (4) $MnO_4^- + H_2O_2 \rightarrow Mn^{2+} + O_{2(g)}$ (acid)

   (5) $Mn^{2+} + HBiO_3 \rightarrow Bi^{3+} + MnO_4^-$ (acid)

   (6) $Zn_{(s)} \rightarrow Zn(OH)_4^{2-} + H_{2\,(g)}$ (basic solution)

RECAP SECTION

The major thrust of Chapter 17 is to give each learner hands-on experience at working with chemical equations and manipulating numbers. If you answered the study-guide questions without difficulty, you have mastered a valuable chemical tool. You are able to take equations apart, analyze them, and make them work for you. You should realize by now that you have all the skills needed to take a word equation and transform it into a balanced tool to be used for chemical description and calculations. You can handle ions, molecules, or gases in equations and calculations. If you have had difficulty, you should go back over the examples in the text very carefully and then come back to the study guide and the review problems at the end of the chapter.

ANSWERS TO QUESTIONS AND SOLUTIONS TO PROBLEMS

1. (1) Hydrogen gas is oxidized and is the reducing agent. Chromium is reduced and is the oxidizing agent.

    $$Cr_2O_3 + 3\ H_2(g) \xrightarrow{\Delta} 2\ Cr + 3\ H_2O$$

    First, assign oxidation numbers. O = −2 and H = +1. Hydrogen gas = 0. Chromium metal = 0.

    $$Cr_2O_3 + 3\ H_2(g) \rightarrow 2\ Cr + 3\ H_2O$$
    $$+3\ -2 \quad\quad 0 \quad\quad\quad 0 \quad\ +1\ -2$$

    What is Cr in Cr_2O_3? If O = −2, then each Cr will have to be +3 in order to balance the total amount of negative oxidation value of 3 oxygen. Which element has been oxidized (lost electrons)? Hydrogen has changed from 0 to +1, so it has been oxidized. Cr has changed from +3 to 0, so it has been reduced.

    (2) Hydrogen gas is oxidized and is the reducing agent. Oxygen gas is reduced and is the ozidizing agent.

    $$2\ H_2(g) + O_2(g) \xrightarrow{\Delta} 2\ H_2O$$
    $$\quad 0 \quad\quad\quad 0 \quad\quad\quad +1\ -2$$

    Each hydrogen atom has lost an electron, and each oxygen atom has gained two electrons.

    (3) Iodine is oxidized and is the reducing agent. Nitrogen is reduced and is the oxidizing agent.

    $$2\ HNO_2(aq) + 2\ HI(aq) \rightarrow I_2(s) + 2\ NO \uparrow +2\ H_2O$$
    $$+1\ +3\ -2 \quad\quad +1\ -1 \quad\quad\ 0 \quad\ +2\ -2 \quad +1\ -2$$

    Iodine changes from −1 to 0, which is a loss of one electron (oxidation). Nitrogen has gained one electron in changing from +3 to +2.

    (4) Sodium is oxidized and hydrogen is reduced. Sodium is the reducing agent, and hydrogen is the oxidizing agent.

    $$2\ Na(s) + H_2O \rightarrow 2\ NaOH + H_2 \uparrow$$
    $$\quad 0 \quad\quad\ +1\ -2 \quad\ +1\ -2\ +1 \quad\ 0$$

Sodium changes from 0 to +1, an oxidation process. Na is the reducing agent. Hydrogen changes from +1 to 0 for one of the products, hydrogen gas. Notice that hydrogen is also present in NaOH at the same oxidation state as in H_2O. Hydrogen is reduced and is the oxidizing agent.

(5) Bromine is both oxidized and reduced.

$$5\ NaBr + NaBrO_3 + 3\ H_2SO_4 \rightarrow 3\ Br_2 + 3\ Na_2SO_4 + 3\ H_2O$$
$$+1-1 \quad +1+5-2 \quad +1+6-2 \quad \quad 0 \quad \quad +1+6-2$$

The only atoms changing oxidation number are Br. Five Br atoms change from −1 to 0. This is oxidation, a loss of electrons. These Br atoms are the reducing agents. The other Br atom changes from +5 to 0, which is a gain of electrons, or reduction. This Br atom is the oxidizing agent.

2. (1) $2\ HNO_2(aq) + 2\ HI(aq) \rightarrow 2\ NO(g) + I_2(s) + 2\ H_2O$

Assign oxidation numbers and write each half-reaction for oxidation and reduction.

$$HNO_2 + HI \rightarrow NO + I_2 + H_2O$$
$$+1+3-2 \quad +1-1 \quad +2-2 \quad 0 \quad +1-2$$

Oxidation $\quad I^- \rightarrow I_2 \quad\quad\quad$ (I^- loses 1 electron)
balance $\quad 2\ I^- \rightarrow I_2 + 2e^-$

Reduction $N^{3+} \rightarrow N^{2+} \quad\quad\quad$ (N^{3+} gains 1 electron)
balance $\quad N^{3+} + 1e^- \rightarrow N^{2+}$

Multiply the reduction equation by 2 to balance the electron gain and loss.

$$2\ I^- \rightarrow I_2 + 2e^-$$
$$2\ N^{3+} + 2e^- \rightarrow 2\ N^{2+}$$

Place each half-reaction along with the coefficient back in the original equation.

$$2\ HNO_2(aq) + 2\ HI(aq) \rightarrow 2\ NO(g) + I_2(s) + H_2O$$

Balance the remaining H and O atoms.

$$2\ HNO_2(aq) + 2\ HI(aq) \rightarrow 2\ NO(g) + I_2(s) + 2\ H_2O$$

(2) $Mn^{4+} + 2\ Cl^-(aq) \rightarrow Cl_2 \uparrow + Mn^{2+}(aq)$

This equation should be fairly simple to balance.

$$Mn^{4+} + Cl^-(aq) \rightarrow Cl_2 \uparrow + Mn^{2+}(aq)$$

Cl^- is oxidized to free elemental chlorine, and Mn^{4+} is reduced to Mn^{2+}.

Oxidation $\quad 2\ Cl^- \rightarrow Cl_2 \uparrow + 2e^- \quad$ (1 electron lost per atom,
$\quad\quad\quad\quad\quad\quad\quad\quad\quad\quad\quad\quad\quad\quad\quad$ 2 electrons lost per molecule)
Reduction $\quad Mn^{4+} + 2e^- \rightarrow Mn^{2+} \quad$ (2 electrons gained per atom)

$$Mn^{4+} + 2\ Cl^-(aq) \rightarrow Cl_2 \uparrow + Mn^{2+}(aq)$$

The equation is now balanced. Check the electrical charges for the last detail.

$$(4+) + (2-) = 2+$$
left side $\quad\quad$ right side

The equation must have the same electrical charge on both sides as well as the same number of atoms of each element.

(3) $ClO_3^-(aq) + 3\ SnO_2^{2-}(aq) \rightarrow Cl^-(aq) + 3\ SnO_3^{2-}(aq)$

Assign oxidation numbers and write each half-reaction for oxidation and reduction.

$ClO_3^-(aq) + SnO_2^{2-}(aq) \rightarrow Cl^-(aq) + SnO_3^{2-}(aq)$
+5−2   +2−2   −1   +4−2

Oxidation $Sn^{2+} \rightarrow Sn^{4+} + 2e^-$ (2e⁻ loss per atom)

Reduction $Cl^{5+} + 6e^- \rightarrow Cl^-$ (6e⁻ gain per atom)

Balance the electron gain and loss first.

$3\ Sn^{2+} \rightarrow 3\ Sn^{4+} + 6e^-$

$Cl^{5+} + 6e^- \rightarrow Cl^-$

Place each coefficient back in the original equation.

$ClO_3^-(aq) + 3\ SnO_2^{2-}(aq) \rightarrow Cl^-(aq) + 3\ SnO_3^{2-}(aq)$

Check whether the atoms of oxygen balance. Do the electrical charges balance also?

(−1) + (6−) = (−1) + (6−)
left side   right side

(4) $4\ NH_3(g) + 5\ O_2(g) \rightarrow 4\ NO(g) + 6\ H_2O(g)$

Assign oxidation numbers and write each half-reaction for oxidation and reduction.

$NH_3(g) + O_2(g) \rightarrow NO(g) + H_2O(g)$
−3+1   0   +2−2   +1−2

This equation has a small wrinkle to it. Oxygen is reduced from 0 to −2 and shows up in two molecules on the product side.

Oxidation $N^{3-} \rightarrow N^{2+} + 5e^-$

Reduction $O_2^0 + 4e^- \rightarrow 2\ O^{2-}$ (from NO and H_2O)

To balance a gain of 4e⁻ and a loss of 5e⁻, we must use 20 as the lowest common denominator.

$4\ N^{3-} \rightarrow 4\ N^{2+} + 20e^-$

$5\ O_2 + 20e^- \rightarrow 10\ O^{2-}$

Put coefficients back in the equation. Remember that we can only have four NO molecules and that the rest of the oxygen on the product side is in water.

$4\ NH_3(g) + 5\ O_2(g) \rightarrow 4\ NO(g) + H_2O(g)$

To balance out the hydrogen, place a 6 in front of the water. The oxygen is now also balanced.

$4\ NH_3(g) + 5\ O_2(g) \rightarrow 4\ NO(g) + 6\ H_2O(g)$

(5) $4\,H^+(aq) + 2\,Br^-(aq) + SO_4^{2-}(aq) \rightarrow Br_2 + SO_2 \uparrow + 2\,H_2O$

Assign oxidation numbers and write each half-reaction for oxidation and reduction.

$H^+(aq) + Br^-(aq) + SO_4^{2-}(aq) \rightarrow Br_2 + SO_2 \uparrow + H_2O$
$+1 -1 +6-2 0 +4-2 +1-2$

Oxidation $\quad 2\,Br^- \rightarrow Br_2 + 2e^-$

Reduction $\quad S^{6+} + 2e^- \rightarrow S^{4+}$

Electron gain and loss is balanced. Insert a 2 in front of Br^-.

$H^+(aq) + 2\,Br^-(aq) + SO_4^{2-}(aq) \rightarrow Br_2 + SO_2 \uparrow + H_2O$

Check for balance of atoms. H and O are still out of balance. We need four H^+ on the left side and two H_2O molecules on the right.

$4\,H^+(aq) + 2\,Br^-(aq) + SO_4^{2-}(aq) \rightarrow Br_2 + SO_2 \uparrow + 2\,H_2O$

Do a check on electrical charges.

$(4+) + (2-) + (2-) = 0$
 left side right side

3. (1) $\quad 4\,Zn(s) + NO_3^- + 10\,H^+ \rightarrow 4\,Zn^{2+} + NH_4^+ + 3\,H_2O$

To begin, write the two half-reactions containing the elements being oxidized and reduced.

oxidation $\quad Zn(s) \rightarrow Zn^{2+}$

reduction $\quad NO_3^- \rightarrow NH_4^+$

The Zn and N are balanced on each side, so now balance the H and O. Remember acid conditions.

$10\,H^+ + NO_3^- \rightarrow NH_4^+ + 3\,H_2O$

Now balance the electrons in the NO_3^- half-reaction.

$8e^- + 10\,H^+ + NO_3^- \rightarrow NH_4^+ + 3\,H_2O$

We now have $8e^-$ available for the oxidation half-reaction of Zn solid. This means we can use four atoms of Zn for each NO_3^-

$4\,Zn(s) \rightarrow 4\,Zn^{2+} + 8e^-$

Adding the two half-reactions together we have

$4\,Zn(s) + NO_3^- + 10\,H^+ \rightarrow 4\,Zn^{2+} + NH_4^+ + 3\,H_2O$

(2) $3\,ClO^- + I^- \rightarrow 3\,Cl^- + IO_3^-$

Write the two half-reactions.

oxidation $\quad I^- \rightarrow IO_3^-$

reduction $\quad ClO^- \rightarrow Cl^-$

The I and Cl are balanced. Now proceed with the H and O using basic conditions.

To balance O and H in the oxidation equation, add $3H_2O$ to the left and $6H^+$ to the right side of the equation.

$$I^- + 3H_2O \rightarrow IO_3^- + 6H^+$$

Add $6OH^-$ to each side

$$6OH^- + I^- + 3H_2O \rightarrow IO_3^- + 6H^+ + 6OH^-$$

Combine $H^+ + OH^- \rightarrow H_2O$

$$6OH^- + I^- + 3H_2O \rightarrow IO_3^- + 6H_2O$$

Rewrite canceling H_2O on each side.

$$6OH^- + I^- \rightarrow IO_3^- + 3H_2O$$

To balance O and H in the reduction equation, add $1\ H_2O$ to the right side of the equation and $2\ H^+$ to the left side.

$$2H^+ + ClO^- \rightarrow Cl^- + H_2O$$

Add $2\ OH^-$ to each side

$$2OH^- + 2H^+ + ClO^- \rightarrow Cl^- + H_2O + 2OH^-$$

Combine $H^+ + OH^- \rightarrow H_2O$

$$2H_2O + ClO^- \rightarrow Cl^- + H_2O + 2OH^-$$

Rewrite canceling H_2O on each side.

$$H_2O + ClO^- \rightarrow Cl^- + 2OH^-$$

Balance each half reaction electrically with electrons.

$$6OH^- + I^- \rightarrow IO_3^- + 3H_2O + 6e^-$$

$$H_2O + ClO^- + 2e^- \rightarrow Cl^- + 2OH^-$$

Equalize loss and gain of electrons. Multiply reduction reaction by 3.

$$6OH^- + I^- \rightarrow IO_3^- + 3H_2O + 6e^-$$

$$3H_2O + 3ClO^- + 6e^- \rightarrow 3Cl^- + 6OH^-$$

Add the two half reactions together, canceling the $6e^-$, $3H_2O$ and $6OH^-$ from each side of the equation.

$$3ClO^- + I^- \rightarrow IO_3^- + 3Cl^-$$

Check: each side of the equation has a charge of -4 and contains the same number of atoms of each element.

4. (1) No reaction. Copper is below lead on the series and therefore will not replace lead ions from solution.

   (2) Reaction. Magnesium is above zinc on the series and will therefore replace zinc ions from solution.

   $$Mg + ZnCl_2 \rightarrow Zn + MgCl_2$$

(3) Reaction. Tin is above hydrogen on the series and will therefore replace hydrogen ions from solution. Reaction will occur.

$$Sn + 2\ HCl \rightarrow SnCl_2 + H_2 \uparrow$$

(4) Reaction. Aluminum is above silver on the series and will replace silver ions from solution.

$$Al + 3\ AgNO_3 \rightarrow Al(NO_3)_3 + 3\ Ag$$

5. The Al^{3+} ions migrate toward the cathode, and the O^{2-} ions migrate toward the anode. To balance the two half-reactions, balance the electron gain and loss.

$$Al^{3+} + 3e^- \rightarrow Al$$
$$O^{2-} + C \rightarrow CO \uparrow + 2e^-$$

If you multiply the Al^{3+} half-reaction by 2 and the O^{2-} half reaction by 3, the electrons will be balanced.

$$2\ Al^{3+} + 6e^- \rightarrow 2\ Al$$
$$3\ O^{2-} + 3\ C \rightarrow 3\ CO \uparrow + 6e^-$$

Added together: $2\ Al^{3+} + 3\ O^{2-} + 3\ C \rightarrow 2\ Al + 3\ CO \uparrow$

6. (1) Molten $NiBr_2$ will exist as the following ions. $NiBr_2 \xrightarrow{heat} Ni^{2+} + 2\ Br^-$ Oxidation, which is a loss of electrons, occurs at the anode. Br^- is the ion capable of losing electrons. The Ni^{2+} ion needs $2e^-$ (reduction) to become a Ni atom.

Therefore, anode reaction $2\ Br^- \rightarrow Br_2 + 2e^-$

The Ni^{2+} ion needs $2e^-$ (reduction) to become a Ni atom.

Cathode reaction $\quad Ni^{2+} + 2e^- \rightarrow Ni^0$
Overall reaction $\quad Ni^{2+} + 2\ Br^- \rightarrow Ni^0 + Br_2$

(2) LiCl will exist as Li^+ and Cl^- ions in the molten state. Oxidation at the anode will involve the Cl^- ion, but $2\ Cl^-$ ions are needed to balance the equation since Cl_2 is produced.

Anode $\quad 2\ Cl^- \rightarrow Cl_2 + 2e^-$

Reduction at the cathode will use the $2e^-$ available to reduce $2\ Li^+$ ions to lithium atoms.

Cathode $\quad 2e^- + 2\ Li^+ \rightarrow 2\ Li^0$
Overall $\quad 2\ Li^+ + 2\ Cl^- \rightarrow 2\ Li^0 + Cl_2$

7. (1) Fe^{2+} loses $1e^-$ to reach the Fe^{3+} oxidation state. Loss of e^- is oxidation, which occurs at the anode. Cr_2^{6+} goes from 6+ to 3+, which is a gain of electrons or reduction. This occurs at the cathode.

(2) In hydrogen peroxide (H_2O_2), oxygen is in a 1− oxidation state whereas in water oxygen is its usual 2−. This is reduction and occurs at the cathode.

I⁻ loses electrons to reach the elemental state or zero oxidation state, I_2. This is oxidation or the anode reaction.

(3) Iron goes from the elemental state to 2+. This involves a loss of two electrons and takes place at the anode. Nickel is reduced from the 3+ state to 2+ state at the cathode.

These examples may be somewhat difficult. Look at them carefully and be sure you are clear on how to determine oxidation states of various atoms. Refer to the text if necessary.

8. (1) voltaic   (2) electrolytic   (3) voltaic

9. Reaction (1) is oxidation, reaction (2) is reduction. Overall reaction:

$$Pb + PbO_2 + 4H^+ \rightarrow 2Pb^{2+} + 2H_2O$$

10. 6.2 g/L

The problem is concerned with calculating the weight of chemical required for a "Normal" concentration solution. This means we have to determine the equivalent weight for $Na_2S_2O_3 \cdot 5H_2O$ for this particular reaction. First we must determine the chemical species in the $Na_2S_2O_3 \cdot 5H_2O$ that changes oxidation number and by how much. Looking at the reaction it appears that sulfur is involved.

Assigning oxidation numbers we find

$$Na_2S_2^{+2}O_3 \rightarrow Na_2S_4^{+2.5}O_6$$

Each sulfur atom "loses" 1/2 e⁻, but we see that each molecule of $Na_2S_2O_3 \cdot 5H_2O$ has 2 S atoms. Therefore, each molecule of $Na_2S_2O_3 \cdot 5H_2O$ loses 1e⁻ which is the same as saying the equivalent weight for this reaction is the same as the molecular weight.

Therefore,

$$0.025 \text{ N solution} = \left(0.025 \, \frac{\cancel{eq. wt.}}{L}\right)\left(248 \, \frac{g}{\cancel{eq. wt.}}\right) = 6 \text{ g/L}$$

11. (1) $MnO_4^- + 5VO^{2+} + H_2O \rightarrow 5VO_2^+ + 2H^+ + Mn^{2+}$

(2) $P_4 + 3H_2O + 3OH^- \rightarrow PH_3 + 3H_2PO_2^-$

(3) $MnO_2 + SO_3^{2-} + H_2O \rightarrow Mn(OH)_2 + SO_4^{2-}$

(4) $5H_2O_2 + 2MnO_4^- + 6H^+ \rightarrow 2Mn^{2+} + 5O_{2(g)} + 8H_2O$

(5) $2Mn^{2+} + 5HBiO_3 + 9H^+ \rightarrow 2MnO_4^- + 5Bi^{3+} + 7H_2O$

(6) $Zn + 2OH^- + 2H_2O \rightarrow Zn(OH)_4^{2-} + H_{2(g)}$

CHAPTER EIGHTEEN
Radioactivity and Nuclear Chemistry

SELF-EVALUATION SECTION

1. Fill in the blank space or circle the correct response.

   In the symbol $^{80}_{27}X$, the atomic number of X is (1) _____ and the mass number of X is (2) _____. The three principle rays or particles coming from the nucleus of a radioactive nuclide are (3) _____, (4) _____, and (5) _____. When a radioactive nuclide loses a beta particle from the nucleus, its atomic number (6) <u>increases/decreases</u> by (7) _____ unit(s). At the same time, its atomic mass is essentially (8) <u>the same/different</u>. When a radioactive nuclide loses a/an (9) _____ particle, its mass number is (10) <u>increased/decreased</u> by 4 amu, and its atomic number is (11) <u>increased/decreased</u> by two units. The time it takes for one-half of a specific amount of a radioactive nuclide to disintegrate is called its (12) _____. $^{223}_{88}Ra$ has a half-life of 11.2 days. If you have 8.0 grams of $^{223}_{88}Ra$ today, in (13) _____ days there will be 1.0 gram of the nuclide remaining.

2. The list below shows various terms related to the common types of radiation. Use it to construct the table as shown.

beta	4 amu	rapid, 90% that of light
gamma	0	photons of light
alpha	1 amu	identical to He^{2+}
$_{-1}^{0}e$	-1	identical to electron
	none	
-1		1/1837 amu
	speed of light	
$_2^4He$	rather slow speed	+2

197

Symbol	Radiation Name	Mass (amu)	Charge	Velocity	Composition

3. Fill in the missing atomic number, atomic mass, or missing symbol for the following reactions. You may need to use a periodic table.

   (1) $^{238}_{92}U \rightarrow {}^{234}Th + {}^{4}_{2}He$

   (2) $_{82}Pb \rightarrow {}^{214}_{83}Bi + {}^{0}_{-1}e$

   (3) $^{210}_{84}Po \rightarrow {}^{206}_{82}Pb + \underline{\hspace{1cm}}$

   (4) $^{209}_{83}Bi + \underline{\hspace{1cm}} \rightarrow {}^{210}_{84}Po + {}^{1}_{0}n$

   (5) $^{35}_{17}Cl + {}^{1}_{0}n \rightarrow \underline{\hspace{1cm}} + {}^{1}_{1}p$

   (6) $\underline{\hspace{1cm}} + {}^{1}_{0}n \rightarrow {}^{24}_{11}Na + {}^{4}_{2}He$

   (7) $^{14}_{7}N + \underline{\hspace{1cm}} \rightarrow {}^{14}_{6}C + {}^{1}_{1}p$

   (8) $^{14}_{7}N + {}^{1}_{0}n \rightarrow {}^{12}_{6}C + \underline{\hspace{1cm}}$

   (9) $^{60}_{28}Ni + {}^{1}_{1}p \rightarrow {}^{57}_{27}Co + \underline{\hspace{1cm}}$

   (10) $^{7}_{3}Li + {}^{1}_{1}p \rightarrow \underline{\hspace{1cm}} + {}^{1}_{0}n$

   Write your answers here.

4. Identify the following reactions as nuclear fission or nuclear fusion reactions.

   (1) $^{2}_{1}H + {}^{2}_{1}H \rightarrow {}^{3}_{2}He + {}^{1}_{0}n$ _____

   (2) $^{235}_{92}U + {}^{1}_{0}n \rightarrow {}^{144}_{56}Ba + {}^{90}_{36}Kr + 2{}^{1}_{0}n$ _____

   (3) $^{7}_{3}Li + {}^{1}_{1}H \rightarrow 2{}^{4}_{2}He$ _____

   (4) $^{2}_{1}H + {}^{2}_{1}H \rightarrow {}^{3}_{1}H + {}^{1}_{1}H$ _____

   (5) ${}^{1}_{0}n + {}^{235}_{92}U \rightarrow {}^{144}_{54}Xe + {}^{90}_{38}Sr + 2{}^{1}_{0}n$ _____

   (6) $^{2}_{1}H + {}^{2}_{1}H \rightarrow {}^{4}_{2}He + {}^{1}_{0}n$ _____

5. Write the following nuclear bombardment reactions using the shorthand notation.

   (1) $^{35}_{17}Cl + {}^{1}_{0}n \rightarrow {}^{35}_{16}S + {}^{1}_{1}p$

   (2) $^{209}_{83}Bi + {}^{2}_{1}H \rightarrow {}^{210}_{84}Po + {}^{1}_{0}n$

(3) $^{14}_{7}N + ^{1}_{0}n \rightarrow ^{12}_{6}C + ^{3}_{1}H$

(4) $^{60}_{28}Ni + ^{1}_{1}p \rightarrow ^{57}_{27}Co + ^{4}_{2}He$

(5) $^{27}_{13}Al + ^{1}_{0}n \rightarrow ^{24}_{11}Na + ^{4}_{2}He$

6. Fill in the blank space or circle the correct response.
High levels of radiation, especially gamma or X rays, are termed (1) _____ radiation. If the dosage is high enough, (2) _____ can occur within several days. The effects of radiation appear to be localized in the (3) _____ of cells. Rapidly growing and dividing cells are (4) most/least susceptible to damage. Long-term or protracted exposure to (5) high/low levels of radiation can lead to health problems at some later time in a person's life. Presently there is concern about the effect that strontium-90, which is chemically similar to the element (6) _____, has on blood cells manufactured in bone marrow. An accumulation of Sr-90 may lead to increased incidence of bone cancer and leukemia. Radiation damage to the nucleus of a cell can affect future generations of a particular species by giving rise to genetic (7) _____. Such events occur when the radiation damages the genetic material—a molecule called (8) _____—but not severely enough to prevent reproduction.

7. Match the names of scientists associated with nuclear chemistry with the appropriate descriptive phrase.

   (1) Becquerel
   (2) Marie Curie
   (3) Rutherford
   (4) E. O. Lawrence
   (5) Hahn and Strassmann

   a. Discovered alpha and beta rays
   b. Developed cyclotron
   c. Coined word "radioactivity"
   d. Found that uranium salts emit rays
   e. Reported first instance of nuclear fission

8. Fill in the blank space or circle the correct response.

   A magnetic field affects the three principal radioactive rays differently. The beta particle, being (1) positively/negatively charged, will be attracted (2) toward/away from the positive plate. The gamma ray, which has a mass of (3) _____ and an electrical charge of (4) _____, (5) will/will not be attracted by the magnetic field. The alpha ray, which has a charge of (6) _____, will be deflected (7) _____ the negative plate.

9. A piece of a wooden tool found at a Northwest Indian fishing village site has been determined to be approximately 10,000 years old. The ^{14}C in the wood has undergone approximately how many half lives of decay?

   Write your answer here.

10. The half life for $^{32}_{15}$P, a common biological radionuclide, is 14.3 days. Starting with 1500 micrograms of $^{32}_{15}$P, how much will you have left after 100 days?

   Do your calculations here.

11. Fill in the blank space or circle the correct response.

   The mass of an atomic nucleus is (1) <u>more/less</u> than the sum of the masses of the particles that make up the nucleus. The difference in mass is known as the

   (2) _____. The energy equivalent to this mass (using the equation E = mc^2) is called the (3) _____ of the nucleus. This amount of energy would be required to (4) <u>put together/pull apart</u> the particles of a particular nucleus. The higher the binding energy, the (5) <u>more/less</u> stable the nucleus is. In both nuclear fission and fusion reactions, the products have less mass than the reactants. The resultant mass losses are accounted for in the very large quantities of energy that are released.

Challenge Problems

12. Calculate the total energy released by the conversion of 1.000 g of matter into energy. The following formulas and information will be useful.

   Energy = mc^2 where m is mass in grams and c = 2.997 × 10^{10} cm/sec, (speed of light)

   $$1 \frac{g \cdot cm^2}{sec^2} = 1 \text{ erg}$$

   1 calorie = 4.184 × 10^7 ergs.

   Do your calculations here.

13. Calculate the binding energy for $^{32}_{16}S$ which occurs naturally at an abundance of 95%. The mass of one atom of $^{32}_{16}S$ is known to be 31.97207 amu (atomic mass units). The masses of the elemental particles are as follows:

   proton = 1.007277 amu   neutron = 1.008665 amu
   electron = 0.0005486 amu.   1 amu = 1.49 × 10^{-3} erg.

   Do your calculations here.

14. The binding energy of one $^{7}_{3}Li$ atom is 6.258 × 10^{-5} erg. Using the masses of the elemental particles listed in problem 12, calculate the actual mass (in amu) of a $^{7}_{3}Li$ atom.

   1 amu = 1.49 × 10^{-3} erg.

   Do your calculations here.

RECAP SECTION

Chapter 18 is a fascinating chapter about a subject that influences us all. We are constantly bombarded by cosmic radiation, nuclear power plants pose problems of disposal of highly radioactive wastes, and we may sometime come into contact with radioactive isotopes during medical treatment. Radioactivity is not something to fear, but it is something to be discussed with knowledge and respect. Every citizen should know the essential details concerning radioactivity and its beneficial uses.

ANSWERS TO QUESTIONS AND SOLUTIONS TO PROBLEMS

1. (1) 27   (2) 80   (3) alpha   (4) beta
   (5) gamma   (6) increases   (7) one   (8) the same
   (9) alpha   (10) decreased   (11) decreased (12) half-life
   (13) 33.6

2. Check your table with Table 18.2 of the text.

3. (1) Atomic number of Th would be 90.
   (2) Mass number of Pb would be 214.
   (3) Missing species would be an alpha particle, ^4_2He.
   (4) Missing species would be a deuterium atom, ^2_1H.
   (5) Missing species would be $^{35}_{16}S$.   (6) Missing species would be $^{27}_{13}Al$.
   (7) Missing species would be ^1_0n.   (8) Missing species would be ^1_1H.
   (9) Missing species would be ^4_2He.   (10) Missing species would be ^7_4Be.

4. (1) Nuclear fusion—two light atoms combine to form a heavier nucleus.
   (2) Nuclear fission—heavy, unstable nucleus splits into two smaller fragments under bombardment with neutrons.
   (3) Nuclear fusion   (4) Nuclear fusion   (5) Nuclear fission

5. (1) $^{35}_{17}Cl(n,p)^{35}_{16}S$
   (2) $^{209}_{83}Bi(^2_1H,n)^{210}_{84}Po$
   (3) $^{14}_{7}N(n,^3_1H)^{12}_{6}C$
   (4) $^{60}_{28}Ni(p,\alpha)^{57}_{27}Co$
   (5) $^{27}_{13}Al(n,\alpha)^{24}_{11}Na$

6. (1) acute   (2) death   (3) nucleus   (4) most
   (5) low   (6) calcium   (7) mutations   (8) DNA

7. (1) d   (2) c   (3) a   (4) b   (5) e

8. (1) negatively   (2) toward   (3) none   (4) zero
   (5) will not   (6) 2$^+$ (7) toward

9. Approximately 2.

   The half-life for $^{14}_{6}C$ is 5668 years.

10. 12 μg

    100 days is 7 half-lives. After this period of time 0.78% of the original material will be left.

    $$0.0078 \times 1500 \, \mu g = 12 \, \mu g$$

11. (1) less   (2) mass defect   (3) binding energy   (4) pull apart
    (5) more

12. 1,000 g $\times$ (2.997 $\times$ 10^{10} cm/sec)^2 = 8.982 $\times$ 10^{20} ergs. Then convert ergs to cal. 2.147 $\times$ 10^{13} cal.

13. Mass defect is equal to 0.29178 amu which is equal to 4.3475 $\times$ 10^{-4} erg/atom of $^{32}_{16}S$.

14. 7.016 amu.

    The calculated mass of the elemental particles in a ^7_3Li atom gives a value of 7.058137 amu. The binding energy is equal to 4.2 $\times$ 10^{-2} amu.

CHAPTER NINETEEN
Chemistry of Some Selected Elements

SELF-EVALUATION SECTION

1. You are given an electrolysis cell with molten LiCl. Give the anode and cathode reactions plus the overall balanced reaction for the production of lithium metal and chlorine gas.

   Write your answers here.

2. A litmus solution turns blue when a small piece of sodium metal is placed in the solution. The metal disappears. Describe what has happened by means of a balanced equation.

   Write your answers here.

3. Balance the following reactions by inspection or electron gain-loss method.

   (1) $Ca + U_2O_5 \xrightarrow{\Delta} U + CaO$

   (2) $Fe_2O_3 + CO \xrightarrow{\Delta} Fe_3O_4 + CO_2$

   (3) $Fe + HNO_3 \longrightarrow Fe(NO_3)_2 + NH_4NO_3 + H_2O$

   (4) $Fe + H_2O \longrightarrow Fe_3O_4 + H_{2(g)}$
   (Hot) (Steam)

   (5) $Al + H_2O \longrightarrow Al_2O_3 + H_{2(g)}$

   (6) $Al + HCl \longrightarrow AlCl_3 + H_{2(g)}$

   (7) $Al(OH)_3 + HCl \longrightarrow AlCl_3 + H_2O$

   Write your answers here.

4. In your own words, describe what is meant by cathodic protection of iron.

Next, from the list of metals given and referring to Table 17.3 in the text, identify by circling those which will give cathodic protection to iron.

    Cr   Al   Pb   Cu   Zn   Sn   Mg

5. Write the formulas for the following common compounds.

  (1) baking soda _____ (6) potash _____
  (2) magnesia _____ (7) limestone _____
  (3) quicklime _____ (8) gypsum _____
  (4) borax _____ (9) Epsom salts _____
  (5) lye _____ (10) soda ash _____

6. List the flame colors for the alkali metal salts indicated.

  (1) Li salts _____
  (2) K salts _____
  (3) Na salts _____

7. What are the two biologically active alkaline earth metals?

8. List the essential materials needed for pig iron production and state the difference between pig iron and steel.

9. The preparation of pig iron from iron ore involves which of the following reactions?

  (1) $Fe + 6HNO_3 \rightarrow Fe(NO_3)_3 + 3NO_{2(g)} + 3H_2O$
  (2) $FeO + CO_{(g)} \rightarrow Fe + CO_{2(g)}$
  (3) $2C + O_{2(g)} \rightarrow 2CO_{(g)}$
  (4) $2Fe + 3Cl_2 \rightarrow 2FeCl_3$
  (5) $Fe_3O_4 + CO_{(g)} \rightarrow 3FeO + CO_{2(g)}$
  (6) $CaO + SiO_2 \rightarrow CaSiO_3$
  (7) $CaCO_3 \rightarrow CaO + CO_{2(g)}$

10. Name the following binary compounds.

   (1) HBr _____   (4) SO$_2$ _____

   (2) OF$_2$ _____   (5) ICl$_3$ _____

   (3) IBr _____   (6) CCl$_4$ _____

11. Draw the Lewis electron-dot structures for the following ions.

   (1) ClO$^-$

   (2) SO$_4^{2-}$

   (3) NO$_3^-$

   (4) PO$_4^{3-}$

12. Determine the oxidation states of the elements in the following compounds.

   (1) ICl$_3$ _____   (5) H$_2$SO$_4$ _____

   (2) HClO$_2$ _____   (6) OF$_2$ _____

   (3) S$_2$Cl$_2$ _____   (7) Cl$_2$O _____

   (4) Cl$_2$O$_6$ _____   (8) BrF$_5$ _____

   Do your calculations here.

13. Balance the following equations and place correct coefficients in front of the chemical formulas.

   (1) H$_2$S + H$_2$SO$_4$ ⟶ S + H$_2$O

   (2) MnO$_2$ + HCl ⟶ Cl$_2$ + MnCl$_2$ + H$_2$O

   (3) SiO$_2$ + HF ⟶ SiF$_4$ + H$_2$O

   (4) KClO$_3$ + HCl ⟶ KCl + H$_2$O + Cl$_2$

(5) $HBr + H_2SO_4 \longrightarrow Br_2 + SO_2 + H_2O$

(6) $HNO_2 + HI \longrightarrow I_2 + NO + H_2O$

(7) $HCl + HNO_3 \longrightarrow Cl_2 + NO + N_2O$

(8) $H_2S + O_2 \longrightarrow SO_2 + H_2O$

Write your answers here.

14. Identify the following reactions of sulfuric acid as either oxidizing agent, acid, or dehydrating agent.

    (1) $NaOH + H_2SO_4 \rightarrow NaHSO_4 + H_2O$ _____

    (2) $C + 2 H_2SO_4 \rightarrow CO_{2(g)} + 2 SO_{2(g)} + 2 H_2O$ _____

    (3) $Cu + 2 H_2SO_4 \rightarrow CuSO_4 + 2 H_2O + SO_{2(g)}$ _____

    (4) $C_{12}H_{22}O_{11} + 11 H_2SO_4 \rightarrow 12 C + 11 H_2SO_4 \cdot H_2O$ _____

    (5) $CaCO_3 + H_2SO_4 \rightarrow CaSO_4 + CO_{2(g)} + H_2O$ _____

15. Below are compounds that are used to prepare oxygen gas. Complete the reactions and balance.

    (1) $HgO \xrightarrow{\Delta}$

    (2) $H_2O \xrightarrow{\text{electrolysis}}$

    (3) $H_2O_2 \longrightarrow$

16. Arrange the halogens (except for astatine) in order of increasing strength as oxidizing agents.

    Write your answer below.

17. Pick two examples of interhalogen compounds from question 12. Which element is considered electropositive?

    Write your answer below.

18. Complete and balance the following equation, which represents a general method for preparing certain halogens.

    $$NaCl + H_2SO_4 + MnO_2 \rightarrow MnSO_4 + Na_2SO_4 + (a) + (b)_{(g)}$$

19. List essential materials needed for sulfuric acid production.

20. The preparation of sulfuric acid from sulfur involves which of the following reactions?

    (1) $2SO_{2(g)} + O_{2(g)} \xrightarrow{V_2O_5} 2SO_{3(g)}$

    (2) $8HI + H_2SO_4 \longrightarrow 4I_2 + H_2S + 4H_2O$

    (3) $C + 2H_2SO_4 \xrightarrow{\Delta} CO_{2(g)} + 2SO_{2(g)} + 2H_2O$

    (4) $H_2SO_4 + SO_{3(g)} \longrightarrow H_2S_2O_7$

    (5) $NaCl + H_2SO_4 \xrightarrow{\Delta} NaHSO_4 + HCl_{(g)}$

    (6) $H_2S_2O_7 + H_2O \longrightarrow 2H_2SO_4$

21. Write formulas for the following common names
    (1) brimstone _____   (5) muriatic acid _____
    (2) fluorspar _____   (6) bleach _____
    (3) Teflon _____   (7) aqua regia _____
    (4) oleum _____   (8) saltpeter _____

CHALLENGE PROBLEMS

22. Aluminum is produced by electrolysis of Al_2O_3. The balanced reaction is

    $$Al_2O_3 + 3\,C \longrightarrow 2\,Al + 3\,CO_{(g)}$$

    How much aluminum is produced when 10 kg of Al_2O_3 is electrolyzed?

    Do your calculations here.

23. The human body is 65.0% oxygen by weight. How many moles of oxygen does a 175 pound person contain?

RECAP SECTION

Chapter 19 introduces the chemistry of important metallic and non-metallic elements. Because metals are a non-renewable resource, we are learning very rapidly that recycling of steel and aluminum products is a necessity. Consequently, the science of metallurgy is expanding into many new areas. The text discusses the chemistry of several of the more important economic metals such as magnesium and aluminum, but all metals produced commercially have interesting chemical and physical properties. And even though the number of nonmetallic elements is small compared with metals, their importance is evidenced by considering that life depends on such elements as carbon, nitrogen, oxygen, sulfur, and phosphorus. The production of sulfuric acid (H_2SO_4) is said to indicate the overall condition of an industrialized economy, and ammonia is of almost equal importance. We will learn more about the biological importance of both nonmetallic and metallic elements in the succeeding chapters.

ANSWERS TO QUESTIONS AND SOLUTIONS TO PROBLEMS

1. Molten LiCl exists as Li^+ ions and Cl^- ions in the molten state. Oxidation (loss of electrons) occurs at the anode and reduction (gain of electrons) occurs at the cathode.

    Cathode $\quad Li^+ + 1e^- \longrightarrow Li^0$

    Anode $\quad 2\ Cl^- \longrightarrow Cl_{2(g)} + 2\ e^-$
    ___
    Balanced overall $\quad 2\ Li^+ + 2\ Cl^- \longrightarrow 2\ Li^0 + Cl_{2(g)}$

2. From the description in the question, a chemical reaction has taken place. The reactants are Na metal and H_2O. Hydroxide ions (OH^-) turn litmus blue. The suggestion is that they are produced from H_2O. The metal disappears and perhaps this means a soluble ion is produced. Let's put these ideas down on paper.

    $$Na^0 + H_2O \longrightarrow Na^+ + OH^-$$

    We have accounted for most of the observed facts. However, notice that the sodium atom has lost an electron and has been oxidized. At the same time, another chemical element must be reduced in order to balance the electron gain and loss. The OH^- ion retains the same oxidation states as found in water. Therefore, we need to look elsewhere. A clue might be that the amount of hydrogen is not balanced. Can we use the hydrogen with an oxidation state of +1 from water to form a hydrogen molecule?

    What must happen to an H^+ ion to be reduced? If two H^+ ions each gain one electron, they will form an H_2 molecule. This will lead us to a balanced equation.

    $$2\ H^+ + 2e^- \longrightarrow H_{2(g)}$$
    $$2\ Na^0 \longrightarrow 2\ Na^+ + 2e^-$$

    Add in the necessary H_2O and OH^- ions

    $$2\ Na^0 + 2\ H_2O \longrightarrow 2\ Na^+ + 2\ OH^- + H_{2(g)}$$

3. (1) $5\ Ca + U_2O_5 \xrightarrow{\Delta} 2\ U + 5\ CaO$
    (2) $3\ Fe_2O_3 + CO \longrightarrow 2\ Fe_3O_4 + CO_2$
    (3) $3\ Fe + 10\ HNO_3 \longrightarrow Fe(NO_3)_2 + NH_4NO_3 + 3\ H_2O$
    (4) $3\ Fe + 4\ H_2O \longrightarrow Fe_3O_4 + 4\ H_{2(g)}$
    (5) $2\ Al + 3\ H_2O \longrightarrow Al_2O_3 + 3H_{2(g)}$
    (6) $2\ Al + 6\ HCl \longrightarrow 2\ AlCl_3 + 3H_{2(g)}$
    (7) $2\ Al(OH)_3 + 6\ HCl \longrightarrow 2\ AlCl_3 + 3\ H_2O$

4. Cathodic protection of iron means using a metal higher on the activity series than iron to stop the iron from corroding or being oxidized. Thus, a metal higher on the list will be oxidized in preference to the iron, which in turn acts as the cathode rather than the anode.

    From the metals given, Cr, Al, Zn, and Mg will serve as anodes when attached to iron.

5. (1) NaHCO₃ (2) MgO (3) CaO
   (4) Na₂B₄O₇ · 10H₂O (5) NaOH (6) K₂CO₃
   (7) CaCO₃ (8) CaSO₄ · 2H₂O (9) MgSO₄ · 7H₂O
   (10) Na₂CO₃

6. (1) red (2) violet (3) yellow

7. Ca and Mg

8. Iron ore, coke, and limestone are needed. Pig iron contains a number of impurities, including more carbon than steel does. Pig iron is converted to steel.

9. (2), (3), (5), (6), (7)

10. (1) hydrogen bromide (4) sulfur dioxide
    (2) oxygen difluoride (5) iodine trichloride
    (3) iodine monobromide (6) carbon tetrachloride

11. (1) $:\!\ddot{Cl}\!:\!\ddot{O}\!:^-$ (2) $\begin{array}{c} :\ddot{O}: \\ :\ddot{O}:S:\ddot{O}: \\ :\ddot{O}: \end{array}^{2-}$

    (3) $:\ddot{O}:^-$ with N bonded to three O's (4) $\begin{array}{c} :\ddot{O}: \\ :\ddot{O}:P:\ddot{O}: \\ :\ddot{O}: \end{array}^{3-}$

    There is one double bond
    and two single bonds between
    the N and the three oxygens.

12. (1) I = +3   (5) H = +1
       Cl = −1   S = +6
                           O = −2

    (2) H = +1   (6) O = +2
       Cl = +3   F = −1
       O = −2

    (3) S = +1   (7) Cl = +1
       Cl = −1   O = −2

    (4) Cl = +6   (8) Br = +5
       O = −2   F = −1

13. (1) 3 H₂S + H₂SO₄ ⟶ 4 S + 4 H₂O
    (2) MnO₂ + 4 HCl ⟶ Cl₂ + MnCl₂ + 2 H₂O
    (3) SiO₂ + 4 HF ⟶ SiF₄ + 2 H₂O
    (4) KClO₃ + 6 HCl ⟶ KCl + 3 H₂O + 3 Cl₂
    (5) 2 HBr + H₂SO₄ ⟶ Br₂ + SO₂ + 2 H₂O
    (6) 2 HNO₂ + 2 HI ⟶ I₂ + 2 NO + 2 H₂O
    (7) 6 HCl + 2 HNO₃ ⟶ 3 Cl₂ + 2 NO + 4 H₂O
    (8) 2 H₂S + 3 O₂ ⟶ 2 SO₂ + 2 H₂O

14. (1) acid (2) oxidizing agent (3) oxidizing agent
    (4) dehydrating agent (5) acid

15. (1) $2HgO \xrightarrow{\Delta} 2Hg + O_{2(g)}$
    (2) $2H_2O \xrightarrow{electrolysis} 2H_{2(g)} + O_{2(g)}$
    (3) $2H_2O_2 \longrightarrow 2H_2O + O_{2(g)}$

16. F_2 is strongest, followed by Cl_2, Br_2, I_2.

17. In compound (1), the I atom is positive; in compound (8), the Br atom is positive. These are the only two interhalogen compounds shown.

18. Compound (a) is H_2O, compound (b) is Cl_2.
    The balanced equation is

    $2NaCl + 2H_2SO_4 + MnO_2 \longrightarrow MnSO_4 + Na_2SO_4 + 2H_2O + Cl_{2(g)}$

19. S, O_2, H_2SO_4, H_2O

20. (1), (4), (6)

21. (1) sulfur(S)   (2) CaF_2 (3) $(C_2F_4)_n$ (4) $H_2S_2O_7$ (5) HCl
    (6) NaOCl   (7) $HCl + HNO_3$ (8) KNO_3

22. The problem is solved in the usual manner. We first find out the number of moles of Al_2O_3 that is represented by 10 kg and then relate this value to the number of moles of product or Al metal.

    10 kg Al_2O_3 = 10,000 g

    formula wt. of Al_2O_3 = 102 $\dfrac{g}{mole}$

Therefore,

$$10,000 \text{ g} \times \dfrac{1 \text{ mole}}{102 \text{ g}} = 98.0 \text{ moles}$$

We know we are using 98.0 moles of Al_2O_3 as a reactant. How many moles of Al metal will we obtain from this amount? The equation states that 1 mole of Al_2O_3 yields 2 moles of Al^0.

$$Al_2O_3 + 3C \longrightarrow 2Al + 3CO_{(g)}$$

Therefore,

$$\dfrac{2 \text{ moles Al}}{1 \text{ mole } Al_2O_3} \times 98.0 \text{ moles } Al_2O_3 = 196 \text{ moles Al}$$

Number of grams = 196 moles Al $\times$ 27.0 $\dfrac{g}{mole}$

$$= 5.29 \times 10^3 \text{ g Al (3 sig. figures)}$$

23. 1.61×10^3 moles of oxygen (O_2)

First convert pounds to grams, determine the mass by percentage and convert to moles.

$$175 \text{ lb} \times \frac{454 \text{ g}}{1 \text{ lb}} \times \frac{65.0\%}{100\%} \times \frac{1 \text{ mol } O^2}{32.0 \text{ g}} = 1.61 \times 10^3 \text{ mol } O^2$$

CHAPTER TWENTY
Organic Chemistry, Saturated Hydrocarbons

SELF-EVALUATION SECTION

1. Fill in the blank space or circle the appropriate response.

   Carbon, with four outer shell electrons, is able to form (1) <u>one/four</u> single (2) <u>ionic/covalent</u> bonds by sharing its electrons with other elements. The bond angles are not planar but describe a (3) <u>cubic/tetrahedronal</u> shape. The remarkable ability of carbon to bond to (4) <u>itself/only metals</u> leads to the possibility of long chain compounds, ring compounds, and structures containing single, double, and even triple bonds. A single bond consists of (5) _____ electrons, a double bond of (6) _____ electrons, and a triple bond of (7) _____ electrons. For example, in the formula C_2H_4, each carbon atom is joined to two hydrogen atoms by a (8) _____ covalent bond, and the two carbon atoms are joined together by a (9) _____ bond. The conventional diagrams of carbon to carbon bonds use two (10) <u>vertical/horizontal</u> dots to represent a single covalent bond, four dots for a double covalent bond and six dots for a triple covalent bond. The four outer shell electrons of carbon ($2s^2 2p^2$) are hybridized during covalent bond formation to form three types of molecular orbitals. The most common type of molecular orbital is known as the sp^3 orbital. Four of these generate a tetrahedronal structure about a central carbon nucleus.

   Since carbon can bond to itself, many organic compounds are "first cousins," or members of a homologous series. Each member of a homologous series differs from the others by a (11) <u>fixed/varying</u> weight. Thus, the alkanes or parrafins differ by only a CH_2 group or a fixed weight of 14.0 amu. Another interesting feature of organic compounds is the regular occurrence of compounds with the same molecular formula but different physical properties, such as boiling points and melting points. For example, the compounds normal butane and isobutane both have the formula C_4H_{10} but have different properties. These compounds are called (12) _____. The existence of isomers produced by the branching of carbon chains is one reason why there are so many organic compounds.

2. Circle the compounds or elements listed that *do not* contain carbon.

   (1) diamond
   (2) sugar
   (3) charcoal
   (4) silicon carbide
   (5) salt
   (6) graphite
   (7) coke
   (8) carbon black
   (9) plastics
   (10) acetic acid
   (11) ammonia
   (12) starch
   (13) carbon dioxide
   (14) sulfuric acid
   (15) fats
   (16) coal

3. Draw the Lewis electron-dot structures of carbon monoxide and carbon dioxide. Keep in mind the possibility of multiple bonds involving more than one pair of electrons.

   Write your answers here.

4. Draw the Lewis electron-dot structures for butane and 3-chloropentane.

   Write your answers here.

214

5. Match the structural formulas and the names for the isomers of hexane. Look for the longest continuous chain of carbon atoms and then the location of the alkyl groups along the chain.

Formulas	Names
(1) $CH_3-CH_2-CH_2-CH_2-CH_2-CH_3$	3-methylpentane
(2) $CH_3-CH_2-CH_2-CH-CH_3$ with CH_3 branch	n-hexane
(3) $CH_3-CH_2-CH-CH_2-CH_3$ with CH_3 branch	2,3-dimethylbutane
(4) $CH_3-CH-CH-CH_3$ with CH_3, CH_3 branches	2-methylpentane
(5) $CH_3-C(CH_3)(CH_3)-CH_2-CH_3$	2,2-dimethylbutane

6. Fill in the missing members of the alkane homologous series. Give the formulas and the names.

Alkanes			
methane	CH_4		
ethane	C_2H_6		
		octane	C_8H_{18}
butane	C_4H_{10}	nonane	C_9H_{20}
pentane	C_5H_{12}		

7. Name the following alkanes and alkyl halides. Look for the longest carbon chain and name them so that the substitutent numbering is as small as possible.

(1) $CH_3-CH_2-CH(CH_3)-CH_3$ _____

(2) $CH_3-CH_2-C(CH_3)(CH_3)-CH(CH_3)-CH_3$ _____

(3) $CH_3-CH_2-CH(Cl)-CH_3$ _____

(4) CH$_3$–CH(Br)–CH$_2$Cl _____

(5) CH$_3$–CH$_2$–CH(I)–CH(CH$_3$)–CH$_3$ _____

(6) (CH$_3$)$_2$CH–CH$_2$Cl _____

(7) CH$_3$–CH(CH$_3$)–CH$_3$ _____

(8) CH$_3$–CH$_2$–C(CH$_3$)$_2$–CH$_2$–CH$_3$ _____

8. Draw the possible structures for the monochlorination of hexane. Eliminate any duplicate structures.

Write your answers here.

9. Either fill in the name or draw the structure for the following cycloalkanes.

Name   Structure

(1) _____   cyclopropane (CH$_2$ with CH$_2$–CH$_2$)

(2) cyclopentane   _____

(3) _____   CH₂—CH₂
                           | |
                           CH₂—CH₂

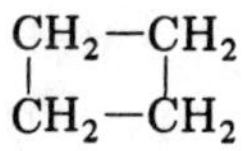

(4) cyclohexane   _____

10. What class of biochemical compounds possesses alkane-like properties?

RECAP SECTION

Chapter 20 has given you an introductory look at the chemistry of hydrocarbons, compounds composed entirely of carbon and hydrogen. In spite of the large number of possible molecules with single and double bonds and branching chains, organic chemists have developed a naming system that allows us to organize our knowledge. Today, organic chemists study reaction mechanisms, molecular structure, and the intricate synthesis schemes involving many types of organic molecules. However, the basic source of carbon compounds remains the hydrocarbons, coal and petroleum. Thus, in addition to being important for fuels, hydrocarbons are the beginning point for the other synthetic organic compounds in our lives.

ANSWERS TO QUESTIONS

1. (1) 4   (2) covalent   (3) tetrahedral   (4) itself
   (5) 2   (6) 4   (7) 6   (8) single
   (9) double   (10) vertical   (11) fixed   (12) isomers

2. Only (5), (11), and (14) do not contain carbon.

3. carbon monoxide   : C ::: O :

   carbon dioxide   : Ö :: C :: Ö :

Carbon has four electrons and an oxygen atom has six. In order to write a correct Lewis electron-dot picture for carbon monoxide, CO, you need to have a triple bond between the two atoms. Two of the bonding electrons are from carbon and four are from oxygen. Carbon dioxide with one carbon atom and two oxygen atoms involves all four of carbon's electrons in forming two double bonds. Each oxygen atom contributes two electrons to form the bonds. As you can see from the electron-dot diagrams for both molecules, each atom has eight electrons around it.

4. butane

```
       H  H  H  H
       ..  ..  ..  ..
    H :C :C :C :C :H
       ..  ..  ..  ..
       H  H  H  H
```

3-chloropentane

```
              ..
       H  H :Cl: H  H
       ..  ..  ..  ..  ..
    H :C :C :C :C :C :H
       ..  ..  ..  ..  ..
       H  H  H  H  H
```

5. (1) n-hexane   (2) 2-methylpentane   (3) 3-methylpentane
   (4) 2,3-dimethylbutane   (5) 2,2-dimethylbutane

6. propane, C_3H_8; hexane, C_6H_{14}; heptane, C_7H_{16}; decane, $C_{10}H_{22}$

7. (1) 2-methylbutane   (2) 2,3,3-trimethylpentane
   (3) 2-chlorobutane   (4) 2-bromo-1-chloropropane
   (5) 3-iodo-2-methylpentane   (6) 1-chloro-2-methylpropane
   (7) 2-methylpropane   (8) 3,3-dimethylpentane

8. There are six carbons in hexane so all the possible structures would be as follows (using a shorthand form):

```
                5    4    3    2
              (Cl) (Cl) (Cl) (Cl)
         6     |    |    |    |        1
       (Cl)—C—C—C—C—C—C—(Cl)
               |    |    |    |    |    |
```

There are six structures to look at for duplicates. Numbers 1 and 6 are identical since the molecule is the same when looked at from either end. Numbers 2 and 5 are identical, as are 3 and 4. Therefore, there are only three possible different monochlorination products of hexane.

$$CH_3-CH_2-CH_2-CH_2-CH_2-CH_2Cl$$
$$CH_3-CH_2-CH_2-CH_2-CHCl-CH_3$$
$$CH_3-CH_2-CH_2-CHCl-CH_2-CH_3$$

9. (1) cyclopropane
   (2)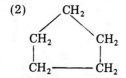
   (3) cyclobutane
   (4)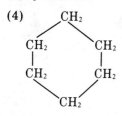
10. Fats and oils.

CHAPTER TWENTY-ONE
Unsaturated Hydrocarbons—Alkenes, Alkynes, and Aromatic Hydrocarbons

SELF-EVALUATION SECTION

1. Fill in the blank space or circle the correct response.

    Alkenes and alkynes are classified as (1) <u>saturated/unsaturated</u> hydrocarbons since their molecules do not contain the maximum number of (2) _____ _____ atoms possible. The alkenes contain at least (3) _____ double bond(s) and the alkynes contain at least (4) _____ triple bond(s) between carbon atoms. The formation of double and triple bonds between carbon atoms is explained by a hybridization process that differs from the sp^3 hybrid orbitals found in alkanes. For alkenes, the hybrid orbitals are termed (5) <u>sp^2/sp</u>. They are formed when one (6) _____ electron is promoted to a (7) _____ orbital to form four half-filled orbitals, $2s^1\ 2px^1\ 2py^1\ 2pz^1$. Three of these orbitals hybridize to form three orbitals in a single plane with angles of (8) _____ between each other. The remaining $2p$ orbital is perpendicular to this plane with a lobe above and below the plane. During double bond formation between two carbon atoms, a hybridized sp^2 orbital from one carbon atom overlaps an identical sp^2 orbital from the second carbon atom. This produces a sigma bond. The two perpendicular orbitals on each carbon atom then overlap to form a pi bond. The pi bond with two electrons has two regions of electron cloud density—one above and one below the sigma bond. The remaining sp^2 orbitals on each carbon can form bonds to hydrogen atoms, halogen atoms, or other carbon atoms, for example. Alkynes with a triple bond have a bond formation process that is similar to that of alkenes. Two equivalent (9) <u>sp^2/sp</u> hybridized orbitals are formed with an angle between them of (10) _____. The remaining unhybridized $2p$ orbitals are oriented at right anbles to each other and the hybridized sp

orbitals. (11) _____ bonds are formed by the hybridized orbitals and two (12) _____ bonds are formed by overlap of the unhybridized p orbitals. In both alkenes and alkynes, the pi bonds are weaker than the sigma bonds, and therefore the molecules are very reactive.

2. Fill in the missing members of the alkene and alkyne homologous series. Give the formulas and the names.

(1)   Alkenes

    no one carbon member

    _____   _____

    propene C_3H_6

    _____   _____

    _____   _____

    hexene C_6H_{12}

    _____   _____

    octene C_8H_{16}

    _____   _____

    decene $C_{10}H_{20}$

(2)   Alkynes

1. no one carbon member
2. ethyne C_2H_2
3. _____   _____
4. _____   _____
5. pentyne C_5H_8
6. _____   _____
7. _____   _____
8. _____   _____
9. nonyne C_9H_{16}
10. _____   _____

3. Name the following compounds.

(1) $CH_2-CH=CH_2$

(2) $CH_3-CH_2-CH_2-C\equiv CH$

(3)
$$CH_2=\underset{\underset{CH_3}{|}}{C}-\underset{\underset{CH_3}{|}}{CH}-CH_3$$

(4)
$$CH_3-\underset{\underset{CH_3}{|}}{CH}-CH_2-CH_2-CH=CH_2$$

(5) $CH_3-CH_2-CH=\underset{\underset{CH_3}{|}}{C}-CH_3$

(6) $CH_3-CH_2-CH=CH_2$

4. Which of the following compounds can exist as geometric isomers?

(1) $\begin{array}{c} CH_3 \\ \diagdown \\ CH_3 \end{array} C=C \begin{array}{c} H \\ \diagup \\ \diagdown H \end{array}$

(2) $CH_3CH_2\underset{\underset{CH_3}{|}}{C}=CH_2$

(3) $\begin{array}{c} CH_3 \\ \diagdown \\ H \end{array} C=C \begin{array}{c} CH_3 \\ \diagup \\ \diagdown H \end{array}$

(4) $CH_2Cl-CH_2-\underset{\underset{CH_3}{|}}{C}=CH_2$

(5) CCl$_2$=CHCl   (6) CHCl=CHCl

5. Draw the geometric structure of the following chemical compounds from their name.

   (1) cis-2-butene

   (2) trans-1,2-dichloroethene

   (3) cis-1-bromobutene

6. Predict the products formed in the following addition reactions. You may have to use Markovnikoff's rule.

   (1) CH≡CH + HCl ⟶

   (2) CH$_2$=CH$_2$ + Br$_2$ ⟶

   (3) CH$_3$−C(CH$_3$)=CH$_2$ + HI ⟶

   (4) CH$_3$−CH$_2$−C(CH$_3$)=CH−CH$_3$ + HBr ⟶

   (5) CH$_3$−CH=CH−CH$_3$ + HI ⟶

   (6) CH$_2$=CH$_2$ + H$_2$O $\xrightarrow{H^+}$

7. Predict the color changes that will be observed during the following reactions.

   (1) CH$_2$=CH$_2$ + Br$_2$ ⟶
       (reddish brown)

   (2) CH$_3$−CH=CH$_2$ + KMnO$_4$ + H$_2$O ⟶
       (purple)

8. Name the aromatic compounds, using an appropriate system. You might choose to use either the numbering system or the o, m, p system if there are two substituents on the benzene ring.

(1)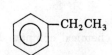

(2)
```
    Cl
   /
  /—Cl
 (benzene ring)
```

_____ _____

(3)
```
    CH3
    |
  (benzene)
    |
    CH3
```

(4)
```
       Br
       |
     (benzene)
    /         \
   Br          Br
```

_____ _____

(5)
```
    CH3
    |
  (benzene)
```

(6)
```
    NO2
    |
  (benzene)—NO2
```

_____ _____

9. Draw the structures for the following compounds.

   (1) bromobenzene _____

   (2) ethylbenzene _____

   (3) phenol _____

   (4) p-bromoaniline _____

   (5) o-chloronitrobenzene _____

   (6) 1,3,5-trinitrobenzene _____

10. Match the correct fused aromatic ring structure with its name.

    (1) napthalene   (a)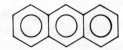

224

(2) anthracene (b)

(3) 3,4-benzpyrene (c)

11. Complete the following reactions that involve aromatic hydrocarbons. Name each reaction type.

(1)

benzene + Cl_2 $\xrightarrow{FeCl_3}$

(2)

benzene + $HO\text{-}NO_2$ ⟶
nitric acid

(3)

benzene + CH_3CHCH_2Cl $\xrightarrow{AlCl_3}$
$\quad\quad\quad\quad\ \ |$
$\quad\quad\quad\quad CH_3$

(4)

⌬—CH_2CH_3 $\xrightarrow[\Delta]{KMnO_4/H_2O}$

RECAP SECTION

Chapter 21 continues the discussion of hydrocarbon molecules begun in Chapter 20. Alkenes and alkynes are considerably more reactive than alkanes by virtue of the double and triple bonds. Aromatic hydrocarbons derived from the parent compound benzene exhibit a reactivity unlike that of the other hydrocarbons. Many important products in everyday use come from unsaturated hydrocarbon molecules. An understanding of these chemicals and their properties should be part of any chemistry student's learning.

ANSWERS TO QUESTIONS

1. (1) unsaturated (2) hydrogen (3) one (4) one
   (5) sp^2 (6) s (7) p (8) 120°
   (9) sp (10) 180° (11) sigma (12) pi

2. (1) ethene, C_2H_4; butene, C_4H_8; pentene, C_5H_{10}; heptene, C_7H_{14}; nonene, C_9H_{18}
   (2) propyne, C_3H_4; butyne, C_4H_6; hexyne, C_6H_{10}; heptyne, C_7H_{12}; octyne, C_8H_{14}; decyne, $C_{10}H_{18}$

3. (1) propene (2) 1-pentyne (3) 2,3-dimethyl-1-butene
   (4) 5-methyl-1-hexene (5) 2-methyl-2-pentene (6) 1-butene

4. The compounds that can exist as geometric isomers are (3) and (6).

5. (1) $\begin{array}{c}CH_3\\ \end{array}\!\!\!\!\!>\!\!C=C\!\!<\!\!\!\!\!\begin{array}{c}CH_3\\ H\end{array}$ with H below left

   (2) $\begin{array}{c}Cl\\ H\end{array}\!\!\!\!\!>\!\!C=C\!\!<\!\!\!\!\!\begin{array}{c}H\\ Cl\end{array}$

   (3) $\begin{array}{c}CH_3-CH_2\\ H\end{array}\!\!\!\!\!>\!\!C=C\!\!<\!\!\!\!\!\begin{array}{c}Br\\ H\end{array}$

6. (1) $CH_2=CHCl$ (2) CH_2Br-CH_2Br

   (3) $CH_3-\underset{I}{\overset{CH_3}{C}}-CH_3$

   (4) $CH_3-CH_2-\underset{Br}{\overset{CH_3}{C}}-CH_2-CH_3$

   (5) $CH_3-CH_2-\underset{I}{CH}-CH_3$ (6) CH_3-CH_2OH

7. (1) The reddish brown will disappear as the bromine adds to the double bond.
   (2) The purple color will become reddish brown. MnO_2 is formed.

8. (1) ethylbenzene (2) ortho-dichlorobenzene or 1,2-dichlorobenzene

(3) para-xylene or
1,4-dimethylbenzene

(4) 1,3,5-tribromobenzene

(5) toluene or methylbenzene

(6) ortho-dinitrobenzene or
1,2-dinitrobenzene

9. (1) bromobenzene (Br on benzene ring)

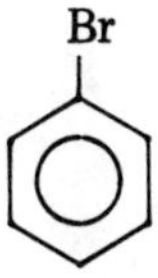

(2) ethylbenzene (CH₂CH₃ on benzene ring)

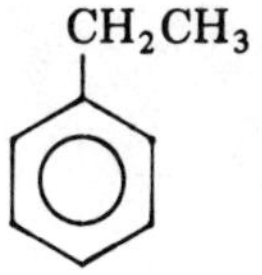

(3) phenol (OH on benzene ring)

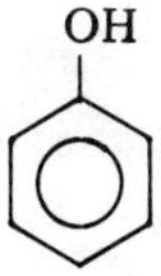

(4) 1,4-dibromo-... (NH₂ and Br para on benzene ring) — para-bromoaniline

(5) 1-chloro-2-nitrobenzene (Cl and NO₂ ortho on benzene ring)

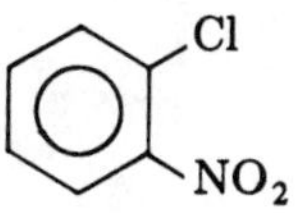

(6) 1,3,5-trinitrobenzene (three NO₂ groups)

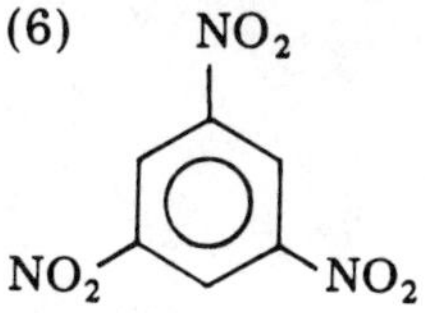

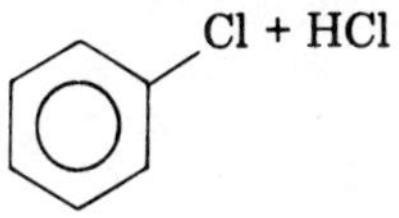

10. (1) c   (2) a   (3) b

11. (1) C₆H₅Cl + HCl — Halogenation

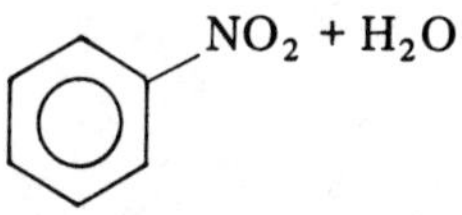

(2) C₆H₅NO₂ + H₂O — Nitration

(3) C₆H₅CH₂CH(CH₃)₂ + HCl — Alkylation (Friedel-Craft)

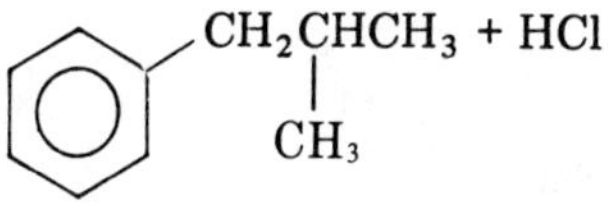

(4) C₆H₅COOH + CO₂ — Oxidation

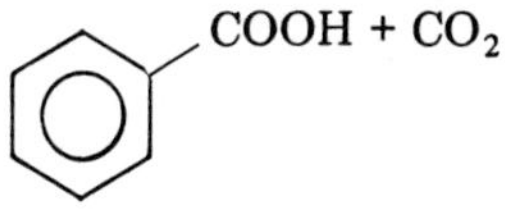

CHAPTER TWENTY-TWO
Alcohols, Phenols, and Ethers

SELF-EVALUATION SECTION

1. Draw the structural formulas for the following alcohols.

   (1) 2-methyl-2-butanol
   (2) 3-chloro-1-butanol
   (3) 2,2-dimethyl-1-propanol
   (4) 3-methyl-2-butanol
   (5) 2,3-dimethyl-1-pentanol

2. Identify the following alcohols as primary, secondary, tertiary, or polyhydroxy.

   (1) 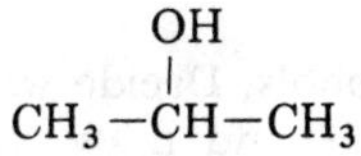
   $$CH_3-\underset{\underset{OH}{|}}{CH}-CH_3$$ _____

   (2) $HO-CH_2-CH_2-OH$ _____

   (3) CH_3-OH _____

   (4) 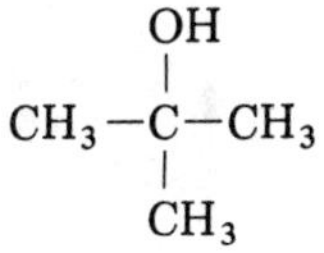
   $$CH_3-\underset{\underset{CH_3}{|}}{\overset{\overset{OH}{|}}{C}}-CH_3$$ _____

   (5) 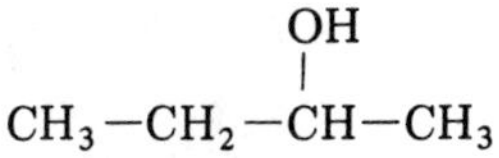
   $$CH_3-CH_2-\underset{\underset{OH}{|}}{CH}-CH_3$$ _____

   (6) 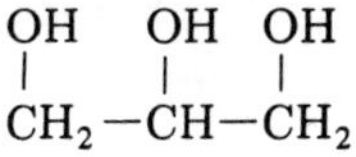
   $$\underset{\underset{OH}{|}}{CH_2}-\underset{\underset{OH}{|}}{CH}-\underset{\underset{OH}{|}}{CH_2}$$ _____

   (7) 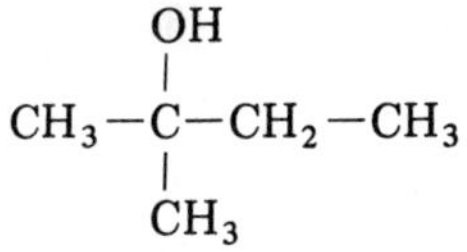
   $$CH_3-\underset{\underset{CH_3}{|}}{\overset{\overset{OH}{|}}{C}}-CH_2-CH_3$$ _____

(8) $CH_3CH_2CH_2CH_2OH$ _____

(9)
$$CH_3-\underset{\underset{CH_3}{|}}{CH}-\underset{\underset{OH}{|}}{\overset{\overset{H}{|}}{C}}-CH_3$$ _____

(10)
$$CH_3-\underset{\underset{CH_3}{|}}{\overset{\overset{CH_3}{|}}{C}}-CH_2OH$$ _____

3. Name the first five alcohols of question 2.

   Write your answers below.

4. Draw all possible isomers of the alcohol with the formula C_4H_9OH and name them.

   Write your answers below.

5. The following section covers various reactions of alcohols. Decide whether a reaction will occur and what the product would be, and draw the chief product molecule and name it.

   (1) Oxidation with dichromate in sulfuric acid

   a. $CH_3CH_2OH \longrightarrow$

   b.
   $$CH_3-\underset{\underset{}{|}}{\overset{\overset{OH}{|}}{CH}}-CH_3 \longrightarrow$$

   c.
   $$CH_3-\underset{\underset{CH_3}{|}}{\overset{\overset{OH}{|}}{C}}-CH_2-CH_3 \longrightarrow$$

   (2) Dehydration with sulfuric acid

   a. $CH_3OH \longrightarrow$

   b.
   $$CH_3-\underset{\underset{}{|}}{\overset{\overset{OH}{|}}{CH}}-CH_3 \longrightarrow$$

c.
$$CH_3-\underset{\underset{CH_3}{|}}{\overset{\overset{OH}{|}}{C}}-CH_2-CH_3 \longrightarrow$$

(3) Synthesis with alkyl halide

$$CH_3-CH_2-CH_2Cl + NaOH(aq) \longrightarrow$$

(4) Reaction of alcohols with sodium metal

$$2\ CH_3-CH_2-CH_2OH + 2\ Na \longrightarrow$$

6. The following are important common alcohols. Draw the structure underneath the name and list all possible descriptive phrases and terms that apply. There is some duplication.

	Terms
(1) Methanol	a. Made from ethylene
	b. Made from propene
	c. Ingredient for beverages
(2) Ethanol	d. Conversion to formaldehyde
	e. Wood alcohol
	f. An industrial solvent
(3) Isopropylalcohol	g. Permanent antifreeze
	h. By-product of soap industry
	i. Synthetic fiber manufacture
(4) Ethylene glycol	j. Used to make explosives
	k. Poisonous
	l. Grain alcohol
(5) Glycerol	m. Used to make acetone
	n. Made from carbon monoxide
	o. From fermentation
	p. Cosmetic manufacture

7. Fill in the blank space or circle the appropriate response.

Ethanol is one of our oldest and best known (1) ketones/alcohols. The preparation by the process known as (2) _____ is still carried out today. The raw materials are usually (3) _____ and (4) _____. A biological catalyst, called a(n) (5) soap/enzyme, is employed in natural fermentation to convert the raw materials into ethanol and carbon dioxide. Ethanol for industrial purposes is made from the two carbon alkene named (6) _____ and sulfuric acid. As a drug, ethanol has been shown to be a (7) depressant/stimulant, which is contrary to many people's beliefs. For nonfood uses, ethanol is made unfit for drinking by a process called (8) _____. This process in effect poisons the ethanol.

8. Give the names for the following compounds.

(1) $CH_3-CH_2-O-CH_2-CH_3$ _____

(2) 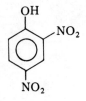 _____

(3) $CH_3-O-CH_2-CH_3$ _____

(4)  _____

(5) _____

(6) $CH_3-CH_2-O-\underset{\underset{CH_3}{|}}{CH}-CH_3$ _____

9. Phenols, which have an —OH group attached directly to a (1) _____ ring, are weak (2) acids/bases. They will react with (3) sodium hydroxide/ sodium bicarbonate solutions but not with (4) _____. In general, phenols are (5) _____ to microorganisms and are (6) _____ to humans if ingested. Phenols used to be widely used as (7) _____ in hospitals, but now the primary use is in the manufacture of (8) _____. Phenol can be derived from coal tar; however, a synthesis involving benzene and propene produces both phenol and (9) _____. The intermediate, isopropylbenzene, is also known as (10) _____.

10. The Williamson synthesis is used to produce ethers. Give the starting sodium alkoxide and alkyl halide needed for the three ethers in question 8.

   Write your answer here.

CHALLENGE PROBLEM

11. Based on the information in problem 22.3 of the text and the definition of molality, calculate the molality of a saturated glucose solution at 20°C.

RECAP SECTION

Chapter 22 has introduced us to some of the various functional groups that occur widely in organic compounds. Alcohols are among the most important compounds from an industrial standpoint and are an everyday part of our lives. Phenols and ethers play important roles also. The interrelationship of hydrocarbons, alcohols, aldehydes, and ketones is further discussed in Chapter 23. Many of the compounds mentioned in Chapters 22 and 23 are involved in biological reactions as well as in industrial processes.

ANSWERS TO QUESTIONS

1. (1)
$$CH_3CH_2\underset{\underset{OH}{|}}{\overset{\overset{CH_3}{|}}{C}}-CH_3$$

   (2)
$$CH_3\underset{\underset{}{}}{\overset{\overset{Cl}{|}}{C}}HCH_2CH_2OH$$

   (3)
$$CH_3\underset{\underset{CH_3}{|}}{\overset{\overset{CH_3}{|}}{C}}CH_2OH$$

   (4)
$$CH_3-\overset{\overset{CH_3}{|}}{C}H-\overset{\overset{OH}{|}}{C}H-CH_3$$

   (5)
$$CH_3-CH_2\overset{\overset{CH_3}{|}}{C}H-\overset{\overset{CH_3}{|}}{C}H-CH_2OH$$

2. (1) secondary   (2) polyhydroxy   (3) primary
   (4) tertiary   (5) secondary   (6) polyhydroxy
   (7) tertiary   (8) primary   (9) secondary
   (10) primary

3. (1) 2-propanol   (2) 1,2-ethanediol or ethylene glycol
   (3) methanol   (4) 2-methyl-2-propanol
   (5) 2-butanol

233

4. There are four possible isomers with a formula of C_4H_9OH:

(1) $CH_3CH_2CH_2CH_2OH$
    1-butanol

(2) $\underset{}{CH_3CH_2\overset{OH}{\underset{|}{C}}HCH_3}$
    2-butanol

(3) $CH_3-\underset{\underset{CH_3}{|}}{\overset{\overset{OH}{|}}{C}}-CH_3$
    2-methyl-2-propanol

(4) $CH_3-\underset{\underset{CH_3}{|}}{CH}-CH_2OH$
    2-methyl-1-propanol

5. (1) a. Primary alcohol oxidized to an aldehyde and carboxylic acid.
       Products: ethanal CH_3CHO and acetic acid CH_3COOH.

   b. Secondary alcohol oxidized to ketone
      Product will be acetone $CH_3-\overset{\overset{O}{\|}}{C}-CH_3$

   c. Tertiary alcohol = no reaction

(2) a. Primary alcohol dehydrates to ether

   CH_3-O-CH_3 dimethyl ether

   b. Secondary alcohol dehydrates to alkene

   $CH_3-CH=CH_2$ propene

   c. Tertiary alcohol dehydrates to alkene

   $CH_3-\underset{\underset{CH_3}{|}}{C}=CH-CH_3$ 2-methyl-2-butene

(3) Product will be an alcohol
    $CH_3CH_2CH_2OH$ propanol

(4) Product will be a sodium alkoxide
    $CH_3CH_2CH_2ONa$ sodium propoxide

6. (1) Methanol   d, e, f, k, n
       CH_3OH

   (2) Ethanol   a, c, l, o, p
       CH_3CH_2OH

   (3) Isopropyl alcohol   b,f,k,m
       $CH_3\underset{\underset{OH}{|}}{CH}-CH_3$

   (4) Ethylene glycol   a, f, g, i, k
       $HO-CH_2-CH_2-OH$

(5) Glycerol   b, h, j, p
HOCH$_2$—CH—CH$_2$OH
              |
              OH

7. (1) alcohols
   (4) sugar
   (7) depressant
   (2) fermentation
   (5) enzyme
   (8) denaturing
   (3) starch
   (6) ethylene

8. (1) diethyl ether
   (2) 2,4-dinitrophenol
   (3) methyl ethyl ether
   (4) *p*-chlorophenol
   (5) *m*-methylphenol
   (6) ethyl isopropyl ether

9. (1) benzene
   (4) sodium bicarbonate
   (7) antiseptics
   (9) acetone
   (2) acids
   (5) toxic
   (8) resins and plastics
   (10) cumene
   (3) sodium hydroxide
   (6) poisonous

10. (1) CH_3CH_2ONa and CH_3CH_2Cl
    (2) CH_3ONa and CH_3CH_2Cl or CH_3CH_2ONa and CH_3Cl
    (3) CH_3CH_2ONa and CH_3—CHCl or CH_3CHONa and CH_3CH_2Cl
                                  | |
                                  CH_3 CH_3

11. 5.3 m   (5.3 $\frac{\text{moles}}{\text{Kg solvent}}$)

CHAPTER TWENTY-THREE
Aldehydes and Ketones

SELF-EVALUATION SECTION

The first three exercises contain material from Chapter 23 as well as earlier ones, and will help you to review alcohols, ethers, etc., as well as add the structural features of aldehydes and ketones to your knowledge.

1. Match the name of the functional group with the generalized formula.

   (1) alkane _____   a. RX
   (2) aldehyde _____   b. $\quad\;\; O$
   $\qquad\qquad\qquad\qquad\qquad\;\; \|$
   $\qquad\qquad\qquad\qquad\; R-C-R$
   (3) alkyl halide _____ c. R—H
   (4) ether _____   d. R—CH=CH$_2$
   (5) ketone _____   e. R—OH
   (6) alcohol _____   f. R—C≡CH
   (7) alkyne _____   g. R—O—R
   (8) alkene _____   h. R—CHO

2. After looking at the following list of structural formulas, match the names with the various formulas. Some of the names will be IUPAC and some will be common names.

   (1) CH$_3$—O—CH$_3$ _____   a. ethanol
   (2) H$_2$C=O _____   b. formaldehyde
   (3) CH$_3$CH$_2$OH _____   c. benzaldehyde
   (4) $\qquad$ O
   $\qquad\quad\;\; \|$
   $\quad\;$ CH$_3$—C—CH$_2$CH$_3$ _____ d. t-butanol
   (5) $\qquad$ OH
   $\qquad\quad\;\; |$
   $\quad\;$ CH$_3$—CH—CH$_3$ _____   e. dimethyl ether
   (6) HOCH$_2$CH$_2$OH _____   f. 1,2-ethanediol or ethylene glycol
   (7) CH$_3$(CH$_2$)$_3$CH$_2$OH _____ g. 3-methyl-2-butanol

237

(8) 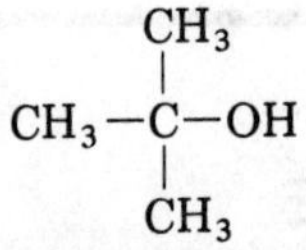   CH₃—C(CH₃)(CH₃)—OH _____   h. butanal

(9) 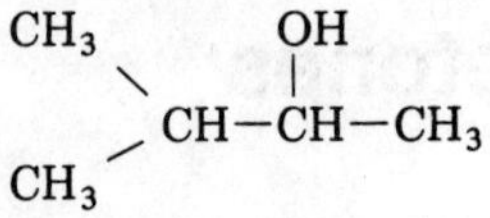   (CH₃)(CH₃)CH—CH(OH)—CH₃ _____   i. 2-methylpropanal

(10) CH₃(CH₂)₂CHO _____   j. 1-pentanol

(11) 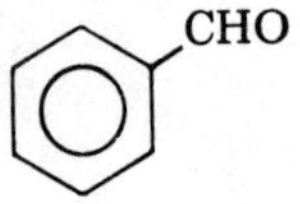 C₆H₅—CHO _____   k. methyl ethyl ketone

(12) 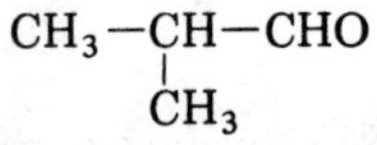 CH₃—CH(CH₃)—CHO _____   l. isopropyl alcohol

3. From the list of formulas given for question 2, identify each one as an alcohol, ether, aldehyde, or ketone.

   Write your answers here.

   (1) _____   (2) _____
   (3) _____   (4) _____
   (5) _____   (6) _____
   (7) _____   (8) _____
   (9) _____   (10) _____
   (11) _____   (12) _____

4. From the list of formulas given for question 2, identify the secondary alcohols and write out the ketone reaction product that would result from an oxidation reaction. Name the ketone product. Example:

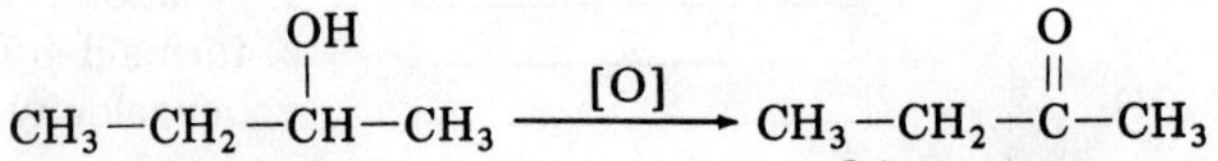

   $$CH_3-CH_2-\underset{\underset{OH}{|}}{CH}-CH_3 \xrightarrow{[O]} CH_3-CH_2-\underset{\underset{O}{||}}{C}-CH_3$$
   2-butanol → 2-butanone or methyl ethyl ketone

   Write your answers here.

5. Write the reaction products for the following reduction reactions of aldehydes and ketones.

   (1) C₆H₅—CHO $\xrightarrow[H_2/Ni]{\Delta}$

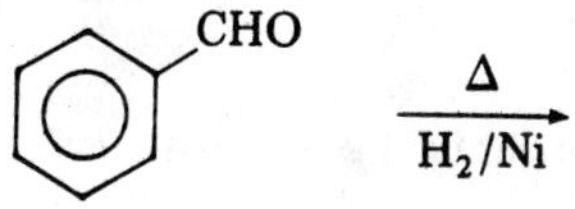

   (2) $CH_3-CH_2-\overset{\overset{O}{\|}}{C}-CH_3$ $\xrightarrow[H_2/Ni]{\Delta}$

   (3) CH_3-CH_2-CHO $\xrightarrow[H_2/Ni]{\Delta}$

6. Aldehydes are easily oxidized to carboxylic acids by mild oxidizing agents. Some of the reactions are the bases of identifying tests. Give the product formed and the appearance for the following tests.

   (1) Tollens' test

   (2) Fehling's test

7. Identify which of the following molecules will undergo aldol condensation and write the products.

   (1) CH_3CHO

   (2) $(CH_3)_2CH-CHO$

   (3) $(CH_3)_3C-\overset{\overset{O}{\|}}{C}-C(CH_3)_3$

   (4) $CH_3-\overset{\overset{O}{\|}}{C}-C(CH_3)_2Cl$

8. Complete the following reactions for the formation of cyanohydrins.

   (1) C₆H₅—CHO + HCN $\xrightarrow{OH^-}$

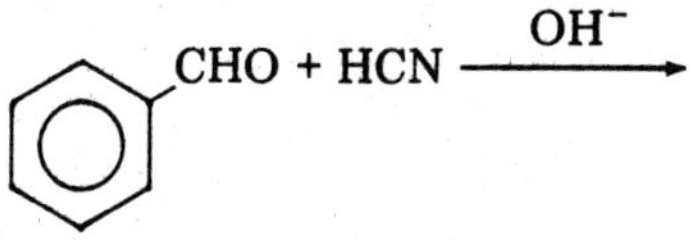

(2) $(CH_3)_2CH-\overset{O}{\underset{\|}{C}}-CH_3 + HCN \xrightarrow{OH^-}$

9. Identify the following molecules as hemiacetals, hemiketals, acetals, or ketals.

(1) $CH_3-CH(OCH_3)_2$

(2) $CH_3-\underset{OH}{\overset{CH_2CH_3}{\underset{|}{\overset{|}{C}}}}-OCH_3$

(3) $CH_3CH_2CH(OH)(OCH_2CH_3)$

(4) $CH_3-\underset{OCH_2CH_3}{\overset{CH_3}{\underset{|}{\overset{|}{C}}}}-OCH_2CH_3$

10. Which of the following molecules will give a positive iodoform test? What is the formula for iodoform?

(1) CH_3-O-CH_3

(2) $H_2C=O$

(3) CH_3CH_2CHO

(4) $CH_3-\overset{O}{\underset{\|}{C}}-CH_2CH_3$

(5) $CH_3-\underset{OH}{\overset{}{\underset{|}{CH}}}-CH_3$

(6) $CH_3-\underset{CH_3}{\overset{CH_3}{\underset{|}{\overset{|}{C}}}}-OH$

(7) $(CH_3)_2CH-\underset{}{\overset{OH}{\underset{}{CH}}}-CH_3$

(8) C_6H_5-CHO

11. Identify from the list given the product alcohol for each Grignard reaction.

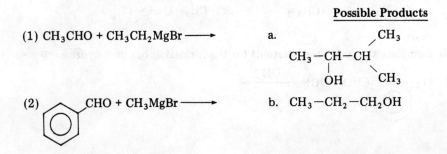

Possible Products

(1) $CH_3CHO + CH_3CH_2MgBr \longrightarrow$ a. $CH_3-CH-CH(CH_3)_2$ with OH

(2) $C_6H_5CHO + CH_3MgBr \longrightarrow$ b. $CH_3-CH_2-CH_2OH$

(3) $CH_3CHO + CH_3-\underset{\underset{MgBr}{|}}{CH}-CH_3 \longrightarrow$

c. $CH_3-\underset{\underset{CH_3}{|}}{\overset{\overset{OH}{|}}{C}}-CH_3$

(4) $CH_3-\overset{\overset{O}{\|}}{C}-CH_3 + CH_3MgBr \longrightarrow$

d. $\underset{}{\text{C}_6\text{H}_5}-\underset{\underset{OH}{|}}{CH}-CH_3$

(5) $CH_2O + CH_3-CH_2MgBr \longrightarrow$

e. $CH_3-\underset{\underset{OH}{|}}{CH}-CH_2-CH_3$

12. What simple chemical test could be used to distinguish between the following pairs of molecules?

(1) CH_3CHO and $CH_3\overset{\overset{O}{\|}}{C}CH_3$

(2) $C_6H_5-CH_2CH_2OH$ and $C_6H_5-\underset{\underset{OH}{|}}{CH}-CH_3$

(3) CH_3CHO and $CH_2=CH-CHO$

RECAP SECTION

Chapter 23 has introduced us to the variety of functional groups possible in organic chemistry. We have seen the importance of the carbonyl group and learned about several different ways in which organic reactions take place. Numerous examples of important industrial and biological chemicals are mentioned in Chapter 23. The chemistry of sugars and carbohydrates is closely related to the concepts discussed in this chapter.

ANSWERS TO QUESTIONS

1. (1) c   (2) h   (3) a   (4) g
   (5) b   (6) e   (7) f   (8) d

2. (1) e (7) j
   (2) b (8) d
   (3) a (9) g
   (4) k (10) h
   (5) l (11) c
   (6) f (12) i

3. (1) ether   (7) alcohol
   (2) aldehyde   (8) alcohol
   (3) alcohol   (9) alcohol
   (4) ketone   (10) aldehyde
   (5) alcohol   (11) aldehyde
   (6) alcohol (diol) (12) aldehyde

4. Secondary alcohols are (5) and (9)

   (5) $\text{CH}_3-\underset{\underset{\text{OH}}{|}}{\text{CH}}-\text{CH}_3 \xrightarrow{[O]} \text{CH}_3-\underset{\underset{\text{O}}{\|}}{\text{C}}-\text{CH}_3$

   2-propanone or acetone

   (9) $\underset{\text{CH}_3}{\overset{\text{CH}_3}{\diagdown}}\text{CH}-\underset{\underset{\text{OH}}{|}}{\text{CH}}-\text{CH}_3 \xrightarrow{[O]} \underset{\text{CH}_3}{\overset{\text{CH}_3}{\diagdown}}\text{CH}-\underset{\underset{\text{O}}{\|}}{\text{C}}-\text{CH}_3$

   3-methyl-2-butanone

5. (1) C₆H₅–CH₂OH (benzyl alcohol)

   (2) $\text{CH}_3-\text{CH}_2-\underset{\underset{\text{OH}}{|}}{\text{CH}}-\text{CH}_3$

   (3) $\text{CH}_3\text{CH}_2\text{CH}_2\text{OH}$

6. (1) Tollens' test—silver mirror produced on test tube. Product is elemental metallic Ag.

   (2) Fehling's test—blue Cu^{2+} ions are reduced to Cu^+ and form brick red precipitate in test tube-Cu_2O-copper(I) oxide.

7. (1), (2), and (4) will undergo aldol condensations.

   (1) $\text{CH}_3\underset{\underset{\text{OH}}{|}}{\text{CH}}\text{CH}_2\text{CHO}$

   (2) $\underset{\text{CH}_3}{\overset{\text{CH}_3}{\diagdown}}\text{CHCH}-\underset{\underset{\text{OH}}{|}}{\overset{\overset{\text{CH}_3}{|}}{\text{C}}}-\text{CHO}$ CH₃

(4)
$$CH_3-\underset{\underset{OH}{|}}{\overset{\underset{|}{CH_3-CCl}}{C}}-CH_2-\underset{\underset{CH_3}{|}}{\overset{O}{\underset{\|}{C}}}-\underset{\underset{CH_3}{|}}{\overset{CH_3}{|}}CCl$$

8. (1)
$$\text{C}_6\text{H}_5-\underset{\underset{OH}{|}}{CHCN}$$

(2)
$$\underset{CH_3}{\overset{CH_3}{\diagdown}}CH-\underset{\underset{CN}{|}}{\overset{\overset{OH}{|}}{C}}-CH_3$$

9. (1) acetal (2) hemiketal (3) hemiacetal
   (4) ketal

10. Positive iodoform tests with the following: (4), (5), (7) must be methyl ketones, secondary alcohols, ethanol, or acetaldehyde. Iodoform is CHI_3.

11. (1) Product is e (2) Product is d
    (3) Product is a (4) Product is c
    (5) Product is b

12. (1) Tollens' test—aldehydes react, ketones don't.
    (2) Iodoform test—secondary alcohols react, primary do not.
    (3) Br_2 addition to double bond.

CHAPTER TWENTY-FOUR
Carboxylic Acids, Their Derivatives, and Amines

SELF-EVALUATION SECTION

1. Name the organic acids using the IUPAC system.

   (1) $CH_3-CH_2-\overset{\displaystyle O}{\underset{\displaystyle OH}{C}}$

   (2) $CH_3-\overset{\displaystyle O}{\underset{\displaystyle OH}{C}}$

   (3) $CH_3-CH_2-\overset{\displaystyle CH_3}{\underset{\displaystyle }{CH}}-\overset{\displaystyle O}{\underset{\displaystyle OH}{C}}$

   (4) Benzoic acid with Cl substituent (C(=O)OH on ring, Cl on ring)

   (5) $Cl-\overset{\displaystyle Cl}{\underset{\displaystyle Cl}{C}}-\overset{\displaystyle O}{\underset{\displaystyle OH}{C}}$

   (6) Benzene ring with C(=O)OH and OH substituents

2. Fill in the blank space or circle the correct response.

   Carboxylic acids show dramatic changes in their physical properties depending upon the number of carbon atoms in the molecule. The carboxyl group end is

245

(1) polar/nonpolar and therefore (2) water soluble/insoluble. The carbon chain portion of the acid is (3) polar/nonpolar and therefore (4) water soluble/insoluble. For acids with one to four carbons, the (5) _____ end of the molecule takes precedence and the compound is (6) water soluble/insoluble. The solubility drops off rapidly after this number of carbons and long-chain fatty acids are insoluble.

3. Fill in the blank spaces with the chemical formula or the name of the following organic acids.

Name	Formula
(1) malonic acid	_____
(2) _____	HOOC—COOH
(3) _____	CH_2=CHCOOH
(4) salicylic acid	_____
(5) lactic acid	_____

4. Predict and name the organic acid product that will result from each reaction.

(1) $CH_3OH + [O] \xrightarrow[H_2SO_4]{Cr_2O_7^{2-}}$ _____

(2) $CH_3CH_2CH_2OH + [O] \xrightarrow[H_2SO_4]{Cr_2O_7^{2-}}$ _____

(3)
$$CH_3-\underset{\underset{CH_3}{|}}{CH}-CH_2OH + [O] \xrightarrow[H_2SO_4]{Cr_2O_7^{2-}}$$ _____

(4)
$$CH_3-\underset{\underset{CH_3}{|}}{CH}-CH_2-CHO + [O] \xrightarrow[H_2SO_4]{Cr_2O_7^{2-}}$$ _____

(5)
$$\underset{CH_3}{\overset{CH_3}{\diagdown}}CH-CH_2OH + [O] \xrightarrow[H_2SO_4]{Cr_2O_7^{2-}}$$ _____

(6) $C_6H_5-CH_2OH + [O] \xrightarrow[H_2SO_4]{Cr_2O_7^{2-}}$ _____

(7) $H_2C=O + [O] \xrightarrow[H_2SO_4]{Cr_2O_7^{2-}}$ _____

(8) 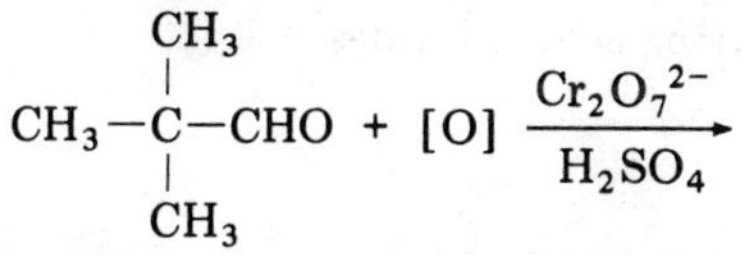 $\text{CH}_3-\underset{\underset{\text{CH}_3}{|}}{\overset{\overset{\text{CH}_3}{|}}{\text{C}}}-\text{CHO} + [\text{O}] \xrightarrow{\text{Cr}_2\text{O}_7^{2-}}{\text{H}_2\text{SO}_4}$ _____

5. Write the structural formulas for:

   (1) Ethyl ethanoate (common name: ethyl acetate)

   (2) n-propyl methanoate (common name: n-propyl formate)

   (3) Methyl benzoate

   (4) Isopropyl ethanoate (common name: isopropyl acetate)

6. The following reactions involve carboxylic acids and their related compounds. Complete the reactions as called for.

   (1) Hydrolysis of a nitrile

   $\text{CH}_3\text{CN} + 2\,\text{H}_2\text{O} \xrightarrow{\text{H}^+}$

   (2) Formation of a salt

   $\text{CH}_3\text{COOH} + \text{NaOH} \longrightarrow$

   (3) Formation of an ester

   $\text{HCOOH} + \text{CH}_3\text{CH}_2\text{OH} \xrightarrow{\text{H}^+}$

   (4) Formation of an amide

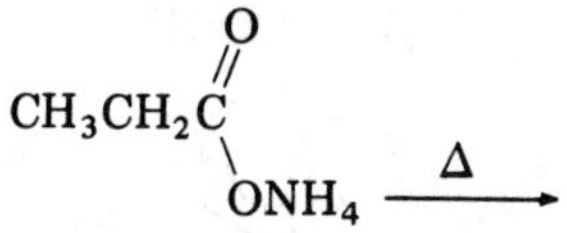

   $\text{CH}_3\text{CH}_2\overset{\overset{\text{O}}{\|}}{\text{C}}\text{—ONH}_4 \xrightarrow{\Delta}$

   (5) Formation of an acid chloride

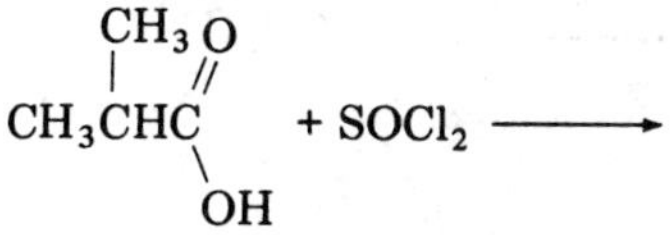

   $\text{CH}_3\underset{|}{\overset{|}{\text{CH}}}\text{—}\overset{\overset{\text{O}}{\|}}{\text{C}}\text{—OH}$ (with CH$_3$ branch) $+ \text{SOCl}_2 \longrightarrow$

7. Complete the following reactions involving acid chlorides.

  (1) Formation of an amide

$$CH_3CH_2\overset{O}{\overset{\|}{C}}-Cl + NH_3 \longrightarrow$$

  (2) Formation of an ester

$$CH_3CH_2\overset{O}{\overset{\|}{C}}-Cl + CH_3OH \longrightarrow$$

  (3) Formation of a carboxylic acid

$$CH_3CH_2\overset{O}{\overset{\|}{C}}-Cl + H_2O \longrightarrow$$

8. Name the following esters, using either the IUPAC system or a common name. Remember to name the acid portion of the ester last.

  (1)
$$CH_3-\underset{O-(CH_2)_4CH_3}{\overset{O}{\overset{\|}{C}}}$$
  _____

  (2)
$$CH_3-(CH_2)_2\underset{O-CH_2CH_3}{\overset{O}{\overset{\|}{C}}}$$
  _____

  (3)
$$H\underset{O-CH_2-CH\underset{CH_3}{\overset{CH_3}{}}}{\overset{O}{\overset{\|}{C}}}$$
  _____

9. Identify the following structural formulas as amines, amides, esters, or acids.

  (1)  OH–C₆H₅ _____ (7) COOH–C₆H₅ _____

  (2) $CH_3-CH(NH_2)-CH_3$ _____ (8) $CH_3\overset{O}{\overset{\|}{C}}-O-(CH_2)_2CH_3$ _____

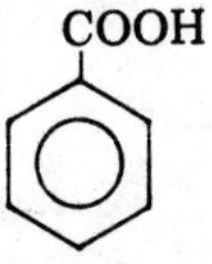

248

(3) 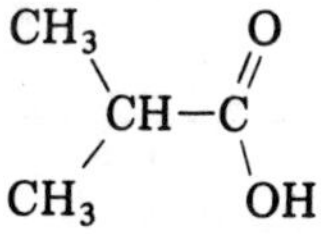 CH_3–$CH(CH_3)$–$C(=O)OH$ ———————

(4) CH_3–NH–CH_3 ———————

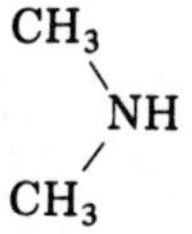

(5) $HC(=O)NH_2$ ———————

(6) C_6H_5–NH_2 ———————

(9) CH_3NH_2 ———————

(10) 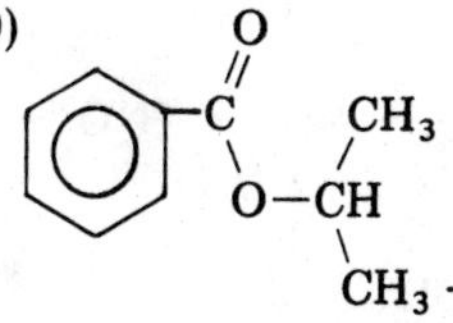 C_6H_5–$C(=O)$–O–$CH(CH_3)_2$ ———————

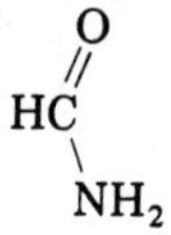

(11) 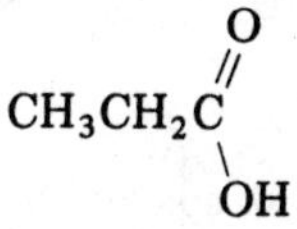 $CH_3CH_2C(=O)OH$ ———————

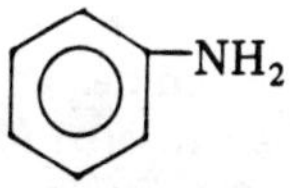

(12) 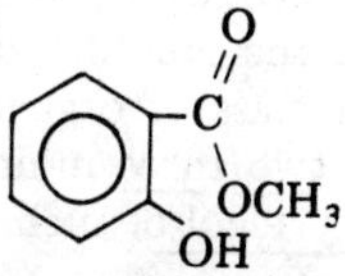 2-hydroxy-C_6H_4–$C(=O)OCH_3$ ———————

10. Examine the following esters and predict what the products will be when each undergoes an alkaline hydrolysis reaction. Name the products. Remember that an acid and an alcohol will result from hydrolysis.

(1) 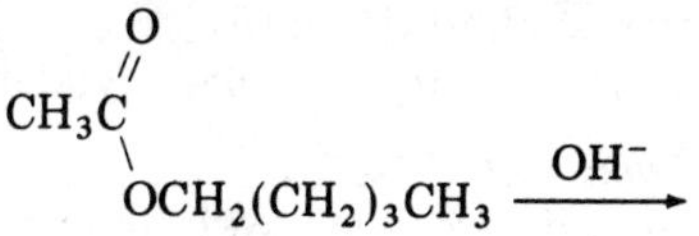 $CH_3C(=O)OCH_2(CH_2)_3CH_3 \xrightarrow{OH^-}$ ———————

(2) $CH_3CH_2CH_2C(=O)OCH_2CH_3 \xrightarrow{OH^-}$ ———————

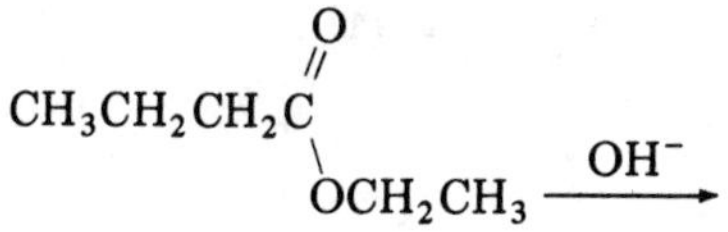

(3) $HC(=O)OCH_2CH(CH_3)CH_3 \xrightarrow{OH^-}$ ———————

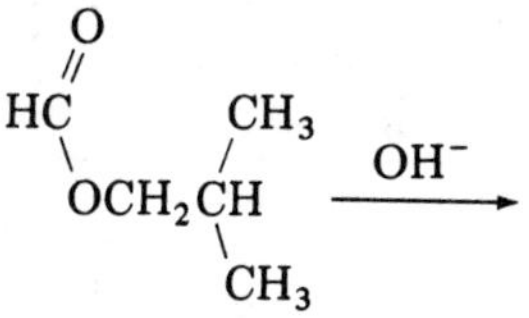

(4) 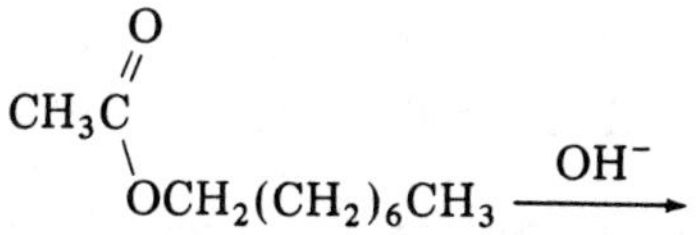 $CH_3C(=O)OCH_2(CH_2)_6CH_3 \xrightarrow{OH^-}$ ———————

11. Write the formulas for the product ions when water reacts with the following amines and acids.

(1) $CH_3NH_2 + H_2O \longrightarrow$ ———————

249

(2) CH_3
　　　　＼
　　　　CH−C(=O)OH　+ H_2O ⟶ _____
　　　　／
　　　CH_3

(3) ⬡−NH_2 + H_2O ⟶ _____

(4) ⬡−OH + H_2O ⟶ _____

12. Fill in the blank space or circle the appropriate response.

    A chemical reaction as important today as when first utilized over 2000 years ago involves the saponification of a fat or oil with (1) acid/base. Fats and oils belong to a class of organic compounds called (2) esters/amides and usually contain (3) two/three molecules of organic acid esterified to a(n) (4) amine/polyhydroxy alcohol such as glycerol. The saponification reaction yields the salts of the organic acids, called (5) _____, and the alcohol. Soaps are detergents and belong to a large class of compounds that have desirable cleansing and lubricating properties.

    Soaps are derived from natural products obtained from animal and plant fats and basic compounds. Synthetic detergents (abbreviated (6) _____) began appearing on the market about 1930. Some of these syndets caused major pollution problems since they were not (7) _____. For some time now, all of the manufactured detergents have been changed to a biodegradable formula. One advantage a syndet has over a natural soap is that positive ions in hard water such as (8) _____ do not form insoluble precipitates with syndets.

    A detergent molecule functions as a cleansing agent by virtue of a nonpolar portion and a polar area in the molecule. A molecule such as sodium lauryl sulfate dissolves in water to produce two ions, (9) _____ and (10) _____. The anion (11) _____ is the detergent species with its hydrocarbon tail, which is (12) nonpolar/polar. The (13) _____ portion dissolves in the greasy substances on soiled clothing, while the (14) _____ end interacts with the water molecules. The detergent species help form a stable emulsion between the oil or grease droplets and the water molecules so that mechanical agitation can lift the dirt from the clothing.

13. Refer to question 8. Give the names of the acid and the alcohol that reacted to produce each of the esters in question 8.

    Write your answer here.

14. Match a term from the left with a description from the right.

    (1) Oleic acid
    (2) Hydrogenolysis
    (3) Saponification
    (4) Linoleic acid
    (5) Hydrogenation
    (6) Linolenic acid
    (7) Hydrolysis

    a. three molecules fatty acid + glycerol
    b. salts of fatty acids + glycerol
    c. yields long chain alcohols + glycerol
    d. one double bond
    e. two double bonds
    f. adds hydrogen to double bonds
    g. three double bonds

15. Give the structure of the molecule that is formed when succinic acid is heated and a molecule of water is removed from the acid.

    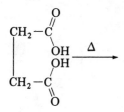

    succinic acid

16. Identify the following amines as primary, secondary, or tertiary, and name them.

    (1) $\begin{array}{c} CH_3 \\ \diagdown \\ N-CH_3 \\ \diagup \\ CH_3 \end{array}$ _____

    (2)   _____

    (3) $\begin{array}{c} H \\ \diagdown \\ N-C_2H_5 \\ \diagup \\ H \end{array}$ _____

(4) 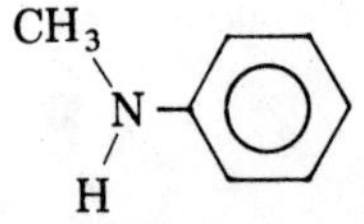 CH$_3$–NH–C$_6$H$_5$

17. Complete the following typical reactions for amines.

   (1) Reaction with an acid

   $$CH_3NH_{2\,(g)} + HCl_{(g)} \longrightarrow$$

   (2) Reaction with acid chlorides

   $$CH_3\overset{O}{\underset{\|}{C}}-Cl + 2\,(CH_3CH_2)_2NH \longrightarrow$$

18. Complete the following reactions used to produce amines from amides and nitriles.

   (1)

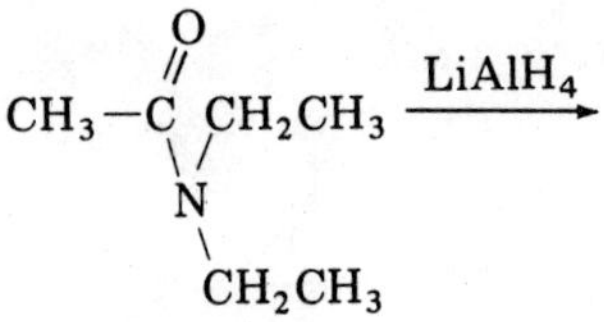

   $$CH_3-\underset{\underset{CH_2CH_3}{\underset{|}{N}}}{\overset{\overset{O}{\|}}{C}}-CH_2CH_3 \xrightarrow{\text{LiAlH}_4}$$

   N,N-diethylacetamide

   (2)

   $$CH_3CH_2C{\equiv}N \xrightarrow{\text{H}_2/\text{Pt}}$$

   propionitrile

RECAP SECTION

Chapter 24 gives further examples of organic chemicals that play important roles in our everyday lives. Paints, soaps, and flavorings are found in every household. The overcoming of the pollution problems brought about by the first syndets, a situation that was not suspected when the detergents came on the market, is an example of how chemical technology can provide solutions to today's problems. The next chapter will discuss other examples of synthetic organic chemicals that have changed the way we live.

ANSWERS TO QUESTIONS

1. (1) propanoic acid (2) ethanoic acid or acetic acid
   (3) 2-methylbutanoic acid (4) m-chlorobenzoic acid
   (5) trichloroethanoic acid (6) p-hydroxybenzoic acid

2. (1) polar (2) water soluble (3) nonpolar
   (4) insoluble (5) polar or carboxyl (6) water soluble

3. (1)
```
COOH
|
CH₂
|
COOH
```
(2) oxalic acid

(3) propenoic acid

(4) 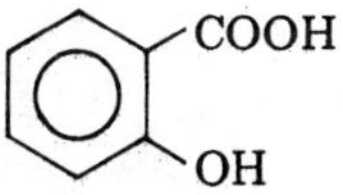 benzene ring with COOH and OH (ortho)

(5) CH₃CHCOOH
       |
       OH

4. (1) methanoic acid, or formic acid   HCOOH

   (2) propanoic acid   CH₃CH₂COOH

   (3) 2-methylpropanoic acid
$$CH_3-\underset{\underset{CH_3}{|}}{CH}-COOH$$

   (4) 3-methylbutanoic acid $CH_3-\underset{\underset{CH_3}{|}}{CH}-CH_2COOH$

   (5) 2-methylpropanoic acid $CH_3-\underset{\underset{CH_3}{|}}{CH}-COOH$

   (6) benzoic acid   C₆H₅—COOH

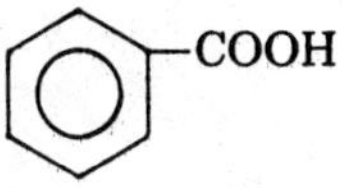

   (7) formic acid or methanoic acid   HCOOH

   (8) 2,2-dimethylpropanoic acid
$$CH_3-\underset{\underset{CH_3}{|}}{\overset{\overset{CH_3}{|}}{C}}-COOH$$

5. (1) 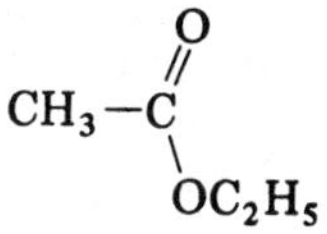
$$CH_3-\underset{}{\overset{O}{\overset{\|}{C}}}-OC_2H_5$$

(2) 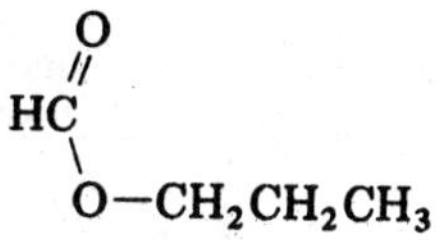
$$HC\underset{}{\overset{O}{\overset{\|}{}}}-O-CH_2CH_2CH_3$$

(3)
$$C_6H_5-\overset{O}{\overset{\|}{C}}-O-CH_3$$

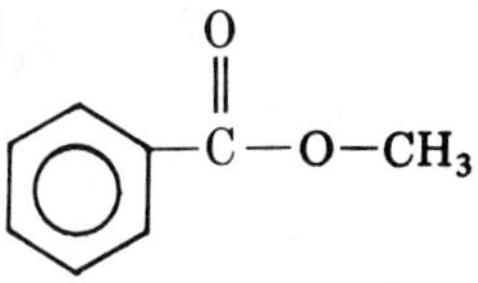

(4) 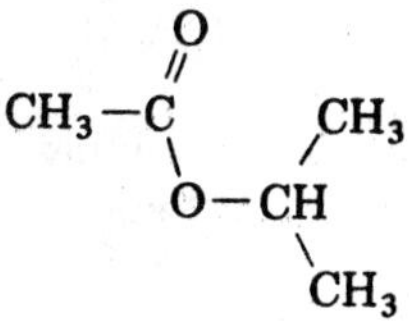
$$CH_3-\overset{O}{\overset{\|}{C}}-O-CH(CH_3)_2$$

6. (1) $CH_3COOH + NH_4^+$

(2) $CH_3\overset{O}{\underset{ONa}{C}} + H_2O$

(3) $H\overset{O}{\underset{O-CH_2-CH_3}{C}} + H_2O$

(4) $CH_3CH_2\overset{O}{\underset{NH_2}{C}} + H_2O$

(5) $CH_3-\underset{Cl}{\overset{CH_3O}{CH}}\overset{}{C} + SO_2 + HCl$

7. (1) $CH_3CH_2\overset{O}{\underset{NH_2}{C}} + NH_4Cl$

(2) $CH_3CH_2\overset{O}{\underset{OCH_3}{C}} + HCl$

(3) $CH_3CH_2\overset{O}{\underset{OH}{C}} + HCl$

8. (1) pentyl ethanoate or pentyl acetate (2) ethyl butanoate
   (3) isobutyl formate or isobutyl methanoate

9. (1) acid   (6) amine   (11) acid
   (2) amine   (7) acid   (12) ester and the OH group
   (3) acid   (8) ester   is an acid
   (4) amine   (9) amine
   (5) amide   (10) ester

10. (1) CH_3COOH acetic acid and $CH_3(CH_2)_3CH_2OH$ 1-pentanol
    (2) $CH_3(CH_2)_2COOH$ butanoic acid and CH_3CH_2OH ethanol
    (3) HCOOH formic acid and $CH_3-\underset{CH_3}{CH}-CH_2OH$ sec-butanol or 2-methyl-1-propanol
    (4) CH_3COOH acetic acid and $CH_3(CH_2)_6CH_2OH$ 1-octanol

11. (1) Product ions will be $CH_3NH_3^+ + OH^-$

(2) $\begin{array}{c} CH_3 \\ \diagdown \\ CH-C \\ \diagup \\ CH_3 \end{array} \begin{array}{c} O \\ \diagup\!\diagup \\ \\ \diagdown \\ O^- \end{array} + H_3O^+$

(3) $C_6H_5\text{-}NH_3^+ + OH^-$ (phenyl ring with NH_3^+)

(4) $C_6H_5\text{-}O^- + H_3O^+$ (phenyl ring with O^-)

12. (1) base   (2) esters   (3) three
    (4) polyhydroxy alcohol   (5) soaps   (6) syndets
    (7) biodegradable   (8) Mg^{2+}, Ca^{2+} (9) Na^+
    (10) lauryl sulfate   (11) lauryl sulfate   (12) nonpolar
    (13) nonpolar   (14) polar

13. (1) acetic acid, 1-pentanol
    (2) butanoic acid, ethanol
    (3) formic acid, isobutyl alcohol

14. (1) d   (2) c   (3) b   (4) e   (5) f   (6) g   (7) a

15.
$$\begin{array}{c} O \\ \| \\ CH_2-C \\ \diagdown \\ O \\ \diagup \\ CH_2-C \\ \| \\ O \end{array}$$ succinic anhydride

16. (1) tertiary, trimethylamine   (2) primary, aniline
    (3) primary, ethylamine   (4) secondary, N-methylaniline

17. (1) $[CH_3NH_3]^+Cl^-$

(2) $\begin{array}{c} O \\ \| \\ CH_3C CH_2-CH_3 \\ \diagdown\diagup \\ N \\ \diagdown \\ CH_2-CH_3 \end{array}$ + $(CH_3CH_2)_2NH_2^+Cl^-$

18. (1) $(CH_3CH_2)_3N$

(2) $CH_3CH_2CH_2NH_2$

CHAPTER TWENTY-FIVE
Polymers–Macromolecules

SELF-EVALUATION SECTION

1. Label the following reactions as chain initiating, chain propagating, or chain terminating.

    (1) 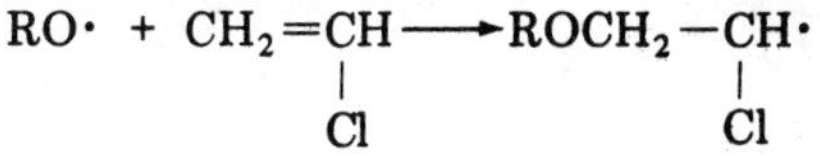 RO· + CH$_2$=CH(Cl) ⟶ ROCH$_2$—CH·(Cl)

    _____

    (2) ROCH$_2$—CH(Cl)—CH$_2$—CH·(Cl) + RO· ⟶ ROCH$_2$—CH(Cl)—CH$_2$—CHOR(Cl)

    _____

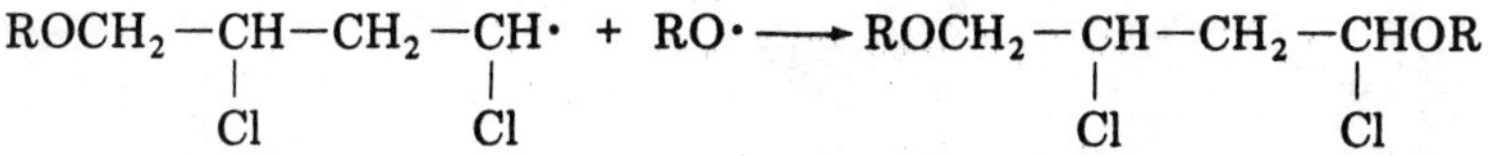

    (3) RO:OR ⟶ 2 RO·

    _____

    (4) 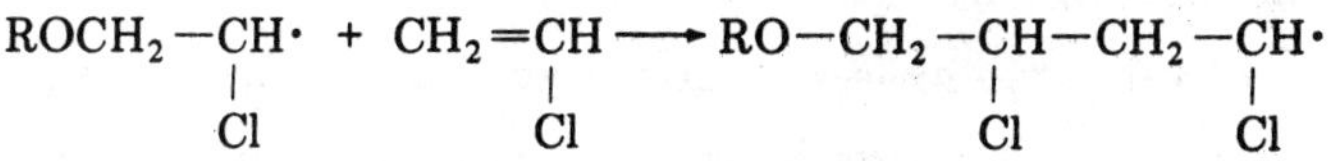 ROCH$_2$—CH·(Cl) + CH$_2$=CH(Cl) ⟶ RO—CH$_2$—CH(Cl)—CH$_2$—CH·(Cl)

    _____

    (5) 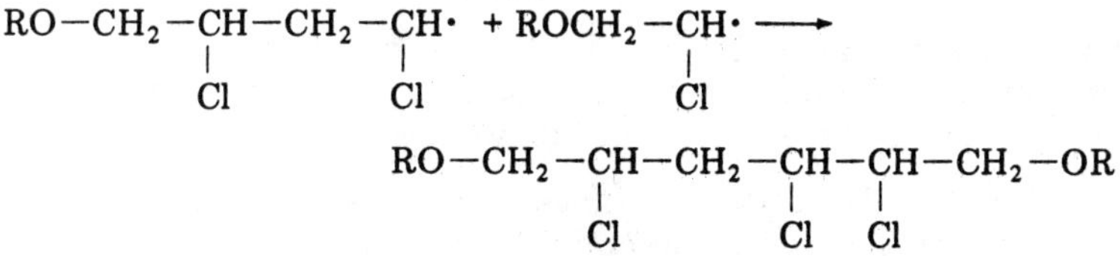 RO—CH$_2$—CH(Cl)—CH$_2$—CH·(Cl) + ROCH$_2$—CH·(Cl) ⟶

    RO—CH$_2$—CH(Cl)—CH$_2$—CH(Cl)—CH(Cl)—CH$_2$—OR

    _____

2. Label the following reactions as either addition or condensation polymerizations.

(1)

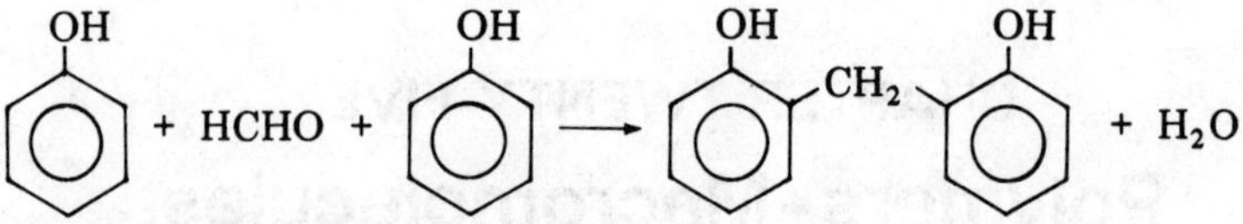

$$\underset{}{\text{C}_6\text{H}_5\text{OH}} + \text{HCHO} + \underset{}{\text{C}_6\text{H}_5\text{OH}} \longrightarrow \text{HO-C}_6\text{H}_4\text{-CH}_2\text{-C}_6\text{H}_4\text{-OH} + \text{H}_2\text{O}$$

_____

(2) $n(\text{CH}_2=\text{CH}) \longrightarrow -(\text{CH}_2-\text{CH})_n-$
         | |
         CN   CN

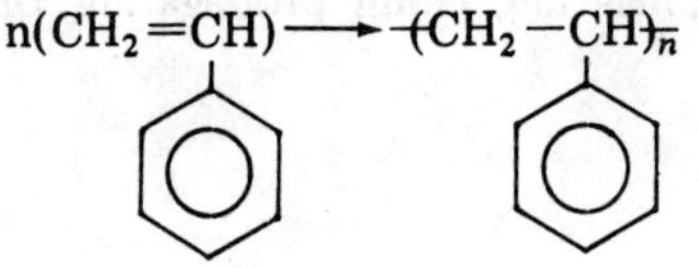

_____

(3)
$$-\underset{\text{OH}}{\overset{\text{O}}{\underset{\|}{\text{C}}}} + \text{H}_2\text{N}-\text{CH}_2- \longrightarrow -(\overset{\text{O}}{\underset{\|}{\text{C}}}-\text{NH}-\text{CH}_2)_n + \text{H}_2\text{O}$$

_____

(4) $n(\text{CH}_2=\text{CH}) \longrightarrow -(\text{CH}_2-\text{CH})_n-$
         | |
       C_6H_5 C_6H_5

_____

(5)
$$-\underset{\text{OH}}{\overset{\text{O}}{\underset{\|}{\text{C}}}} + \text{HO}-\text{CH}_2- \longrightarrow -(\overset{\text{O}}{\underset{\|}{\text{C}}}-\text{O}-\text{CH}_2)_n + \text{H}_2\text{O}$$

_____

3. Fill in the blank space or circle the appropriate response.

The formation of very large molecules from smaller units is a process known as (1) _____ . The large molecule is called the (2) _____ and the small unit, the (3) _____ . For the polymer, polystyrene, the monomer unit is (4) _____ . Another term often used in reference to polymers is (5) micromolecule/macromolecule. The first synthetic polymers were made in the early part of the twentieth century. One particular one, discovered by Leo Baekeland in 1909 and called (6) _____ , was the first commercially successful polymer. It is still used for many applications such as electrical insulators and automobile parts. There are various ways to classify polymers. One way is based on the effect of temperature on a particular polymer. If the polymer softens on heating and can be reformed, it is said to be (7) _____ . On the other hand, a polymer that is hard, brittle, and will not soften when heated is a (8) _____ polymer. In these

258

types of plastics, the polymeric chains are joined together or (9) _____ _____ to form a network structure.

Another classification is based on the manner in which the monomer units join together. When the monomers join together by a free radical reaction mechanism and produce a new longer chain-free radical, the polymer is called a(n) (10) _____ polymer. Other monomers have two reactive sites and join together with the splitting out of a small molecule such as H_2O. The chain can grow at either end. This type of polymer is called a(n) (11) _____ polymer. Polyethylene, PVC, Orlon, Teflon, and Lucite are all examples of (12) _____ polymers, while Dacron, Nylon, Bakelite, and polyurethanes are examples of (13) _____ polymers. The addition polymer, natural rubber, is usually vulcanized during its manufacture into various products. The process invented by (14) _____ in 1839 introduces (15) _____ atoms that (16) _____ the long polymeric chains.

4. Match the monomer molecules listed in the left-hand column with the generalized formulas for the corresponding polymers in the right-hand column. There may be more than one monomer associated with a polymer, as in a copolymer.

(1) $CH_2=C(Cl)(Cl)$

(2) $CH_2=CH-CN$

(3) $CH_3-O-\overset{O}{\underset{\|}{C}}-\langle\bigcirc\rangle-\overset{O}{\underset{\|}{C}}-OCH_3$

(4) $CH_2=C(CH_3)-\overset{O}{\underset{\|}{C}}-OCH_3$

(5) $CH_2=CH-\bigcirc$

a. $+CF_2-CF_2+_n$

b. $\left(CH_2-CH(\bigcirc)\right)_n$

c. $+CH_2-CCl_2+_n$ with Cl, Cl

d. $\left(CH_2-\underset{COOCH_3}{\overset{CH_3}{C}}\right)_n$

e. $\left(CH_2-\underset{CH_3}{\overset{H}{C}}\right)_n$

(6) $CF_2=CF_2$ _____

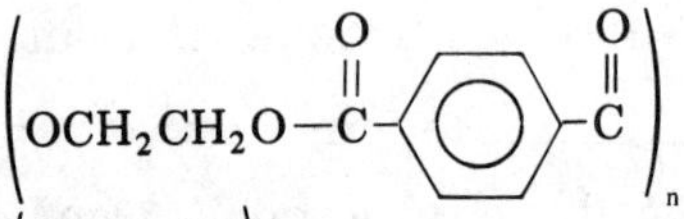

(7) 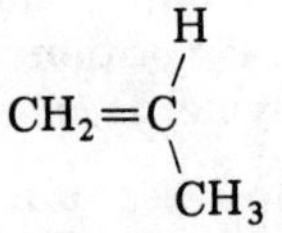 $CH_2=C\overset{H}{\underset{CH_3}{}}$ _____

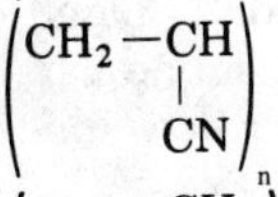

(8) $HOCH_2CH_2OH$ _____

f. $\left(OCH_2CH_2O-\overset{O}{\underset{\|}{C}}-\underset{}{\bigcirc}-\overset{O}{\underset{\|}{C}} \right)_n$

g. $\left(CH_2-\underset{CN}{CH} \right)_n$

h. 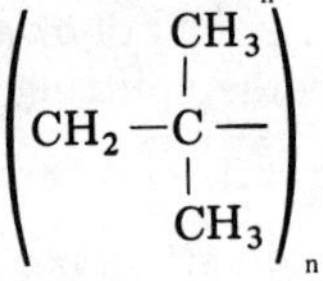 $\left(CH_2-\underset{CH_3}{\overset{CH_3}{\underset{|}{C}}}- \right)_n$

5. Identify which of the following structures is the *cis* configuration for natural rubber and which is the *trans* configuration.

(1) 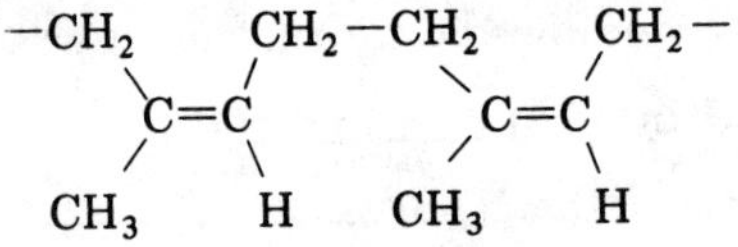 $-CH_2\underset{CH_3}{\overset{}{\diagdown}}C=C\overset{CH_2-CH_2}{\underset{H}{\diagup}}\quad \underset{CH_3}{\overset{}{\diagdown}}C=C\overset{CH_2-}{\underset{H}{\diagup}}$

(2) $-CH_2\underset{CH_3}{\overset{}{\diagdown}}C=C\overset{H}{\underset{CH_2-CH_2}{\diagup}}\quad \underset{CH_3}{\overset{}{\diagdown}}C=C\overset{CH_2-}{\underset{H}{\diagup}}$

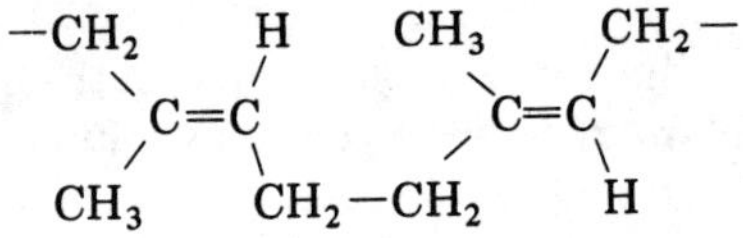

RECAP SECTION

Chapter 25 deals with the topic of polymers, or plastics, as we commonly call them. It has been proposed that historians term the twentieth century the "Age of Plastics" in the same manner that earlier ages are called the Stone Age and the Iron Age. The influence that polymers have had on modern living is tremendous and pervasive. They are involved in everything we do from birth to death.

ANSWERS TO QUESTIONS

1. (1) chain propagating
   (2) chain terminating
   (3) chain initiating
   (4) chain propagating
   (5) chain terminating

2. (1) condensation
   (4) addition
   (2) addition
   (5) condensation
   (3) condensation

3. (1) polymerization
   (4) styrene
   (7) thermoplastic
   (10) addition
   (13) condensation
   (16) cross-link
   (2) polymer
   (5) macromolecule
   (8) thermosetting
   (11) condensation
   (14) Charles Goodyear
   (3) monomer
   (6) Bakelite
   (9) cross-linked
   (12) addition
   (15) sulfur

4. (1) c   (2) g   (3) f
   (4) d   (5) b   (6) a
   (7) e   (8) f

5. (1) in *cis* form   (2) in *trans* form

CHAPTER TWENTY-SIX
Stereoisomerism

SELF-EVALUATION SECTION

1. Examine the following molecules. Determine if there are asymmetric carbon atoms present and indicate their location.

   (1) 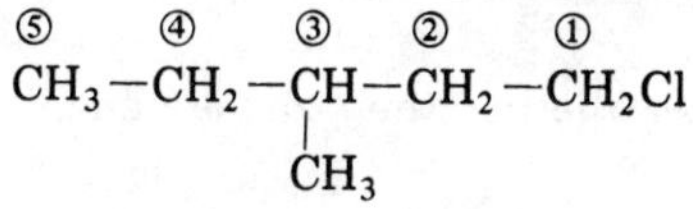 ⑤   ④   ③   ②   ①
   $CH_3-CH_2-CH-CH_2-CH_2Cl$
                 $|$
                CH_3 _____

   (2)   ④   ③ CH_3 ②   ①
                $|$
        $CH_3-C-CH_2-CH_3$
                $|$
                CH_3 _____

   (3) ③   ② ① O
                    $\parallel$
      CH_3-CH_2-C
                      $\backslash$
                      OH _____

   (4) ④   ③   ②   ①
      $CH_3-CHCl-CH_2-CH_3$ _____

   (5) ③   ② ① O
                   $\parallel$
      CH_3-CH-C
            $|$ $\backslash$
           NH_2 OH _____

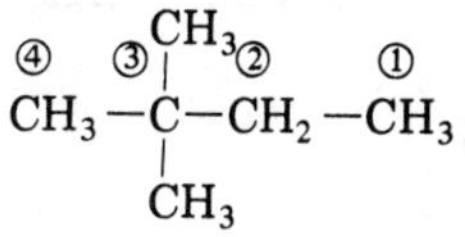

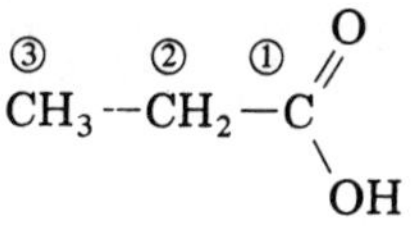

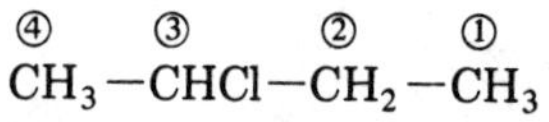

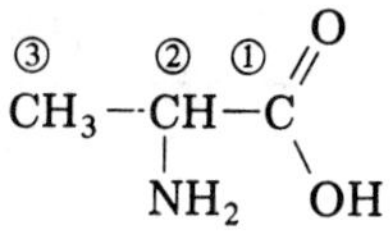

(6) ④   ③   ②   ①
    CH₃−CHCl−CH−CH₃   _____
                |
               CH₃

(7) ③   ②   ①
    CH₂Cl−CHCl−CHClBr   _____

(8)   H ① O
        \ ‖
         C
      ② |
      H−C−OH
      ③ |
      H−C−OH   _____
      ④ |
      HO−C−H
      ⑤ |
      CH₂OH

2. Examine the pairs of molecules. Indicate whether the pairs are enantiomers, diastereomers, or neither.

(1)   H   H
             | |
        CH₃ ─┼─ COOH   and   HOOC ─┼─ CH₃   _____
             NH₂   NH₂

(2)   CH₃   CH₃
             | |
       H₃C ─┼─ CH₃   and   H₃C ─┼─ CH₃   _____
             C₂H₅   C₂H₅

(3)   CH₃   CH₃
             | |
          H−C−Br   and   H−C−Br   _____
             | |
          H−C−Cl   Cl−C−H
             | |
            CH₃   CH₃

(4)   CH₃   CH₃
             | |
          H−C−Br   and   Br−C−H   _____
             | |
          Cl−C−H   H−C−Cl
             | |
            CH₃   CH₃

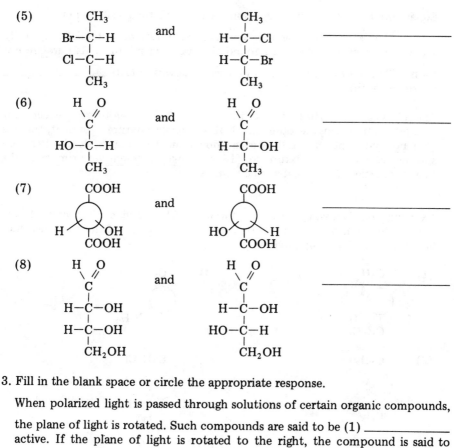

3. Fill in the blank space or circle the appropriate response.

When polarized light is passed through solutions of certain organic compounds, the plane of light is rotated. Such compounds are said to be (1) _____ active. If the plane of light is rotated to the right, the compound is said to be (2) _____, and if the rotation is to the left, the compound is termed (3) _____. The famous French scientist Louis (4) _____ is credited with discovering that crystals of an organic salt, sodium ammonium tartrate, existed in two distinct forms. When he separated the two forms he found that either kind of crystal would rotate plane polarized light, but in opposite directions.

In 1874 it was concluded that in the molecule of an optically active substance the key factor is the presence of a(n) 5 <u>asymmetric/dissimilar</u> carbon atom. Such a carbon atom has (6) <u>three/four</u> different (7) _____ attached to it. Optically active isomers and (8) _____ isomers are examples of stereoisomers.

When two stereoisomers are related to each other as your left hand is to your right hand, the isomers are (9) _____ images of each other. The technical term is (10) <u>enantiomer/optimer</u>.

Equal amounts of mirror image isomers form mixtures called (11) _____

_____ mixtures. During the biological synthesis of potentially optically active compounds, it has been found that racemic mixtures (12) <u>are/are not</u> formed. This result suggests that biological catalysts or (13) _____ are stereospecific.

The physical properties of enantiomers are (14) <u>alike/different</u> except for optical rotation. On the other hand, if a racemic mixture is present, there is (15) <u>no/some</u> optical rotation even though melting points, solubilities, and specific gravity of the mixture are (16) <u>strikingly different/very similar</u> to the same properties in the isolated enantiomers.

4. Examine the following structural formulas. The open circle represents the possible location of an asymmetric carbon atom. Which of the formulas show an optically active compound?

(1) C_3H_7 / H, H, CH_2Cl

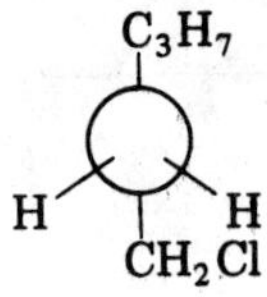

(2) C_3H_7 / Cl, H, CH_3

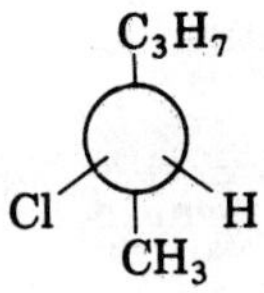

(3) C_2H_5 / CH_3, Cl, C_2H_5

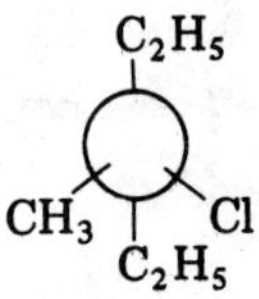

(4) C_3H_7 / Cl, CH_3, CH_3

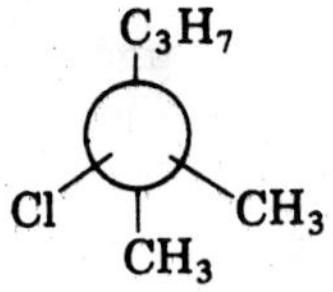

(5) C_3H_6Cl / Cl, H, CH_3

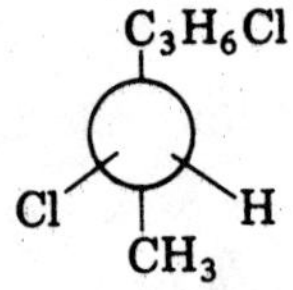

(6) C_3H_7 / H, Cl, CH_3

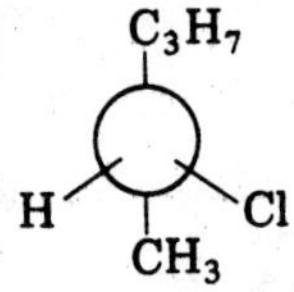

Write your answers here.

5. Examine the following two molecules. Is either one a chiral molecule? If so, draw a projection formula of the enantiomers.

   (1) $CH_3CHClCH_2CH_3$ (2) CH_3CH_2COOH

   Do your drawing here.

6. Using the information in question 1, calculate the number of stereoisomers for each chiral molecule.

   Do your calculations here.

7. Examine the following molecule. Determine whether it is chiral and if so how many asymmetric carbon atoms there are. Then draw the mirror image pairs and determine if they are superimposable. Label the enantiomers, diastereomers, and meso forms, if any.

   $CH_2Cl-CHCl-CHClBr$

8. Can either of the following compounds exist as meso forms? Draw projection formulas to be sure of your answer.

   (1) 2-bromobutane (2) 2,4-dibromopentane

   Draw your projection formulas here.

RECAP SECTION

Chapter 26 has presented a topic of unique importance to organic chemistry—stereoisomerism. The insights an understanding of this subject affords the scientist are numerous. As an example, recall that living systems synthesize and utilize only one enantiomer if an asymmetric carbon atom exists in a molecule. Therefore, various drugs that are manufactured as racemic mixtures are only utilized to half their total concentration by living systems. Only one enantiomer of a particular amino acid is found to be incorporated into proteins. The introductory material in the chapter, even though it is brief, has indicated some of the complexities involved with studying stereoisomers.

ANSWERS TO QUESTIONS

1. (1) Carbon number 3 is asymmetric. Carbon number 1 is not, because there are two hydrogens attached.
   (2) There are no asymmetric carbon atoms.
   (3) There are no asymmetric carbon atoms.
   (4) Carbon number 3 is asymmetric.
   (5) Carbon number 2 is asymmetric.
   (6) Carbon number 3 is asymmetric. Carbon number 2 is not, because it has two methyl groups attached.
   (7) Carbon number 1 and carbon number 2 are asymmetric.
   (8) Carbons number 2, 3, and 4 are asymmetric.

2. (1) enantiomers
   (2) neither (there are 3 CH_3 groups attached to the central carbon atom, hence no asymmetry)
   (3) diastereomers
   (4) enantiomers
   (5) neither; they are identical molecules
   (6) enantiomers
   (7) neither; there are two identical COOH groups
   (8) diastereomers

3. (1) optically   (2) dextrorotatory   (3) levorotatory   (4) Pasteur
   (5) asymmetric   (6) four   (7) atoms or groups   (8) geometric
   (9) mirror   (10) enantiomer   (11) racemic   (12) are not
   (13) enzymes   (14) alike   (15) no   (16) strikingly different

4. (2), (5), and (6) are optically active. The other compounds do not have four different atoms or groups attached to the possible asymmetric carbon atom.

5. (1) is chiral, (2) is not. The enantiomer projection formulas would be

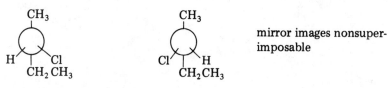

mirror images nonsuperimposable

6. Using the molecules in question 1, we found that
   (1) has one asymmetric carbon atom. Therefore,
   $2^1 = 2$ stereoisomers
   (4) $2^1 = 2$ stereoisomers
   (5) $2^1 = 2$ stereoisomers
   (6) $2^1 = 2$ stereoisomers
   (7) $2^2 = 4$ stereoisomers
   (8) $2^3 = 8$ stereoisomers

7. The molecule is chiral and has two asymmetric carbon atoms. Mirror images would be as follows:

$$
\begin{array}{cccc}
\text{CH}_2\text{Cl} & \text{CH}_2\text{Cl} & \text{CH}_2\text{Cl} & \text{CH}_2\text{Cl} \\
\text{Cl}-\overset{|}{\text{C}}-\text{H} & \text{H}-\overset{|}{\text{C}}-\text{Cl} & \text{Cl}-\overset{|}{\text{C}}-\text{H} & \text{H}-\overset{|}{\text{C}}-\text{Cl} \\
\text{Cl}-\overset{|}{\text{C}}-\text{H} & \text{H}-\overset{|}{\text{C}}-\text{Cl} & \text{H}-\overset{|}{\text{C}}-\text{Cl} & \text{Cl}-\overset{|}{\text{C}}-\text{H} \\
\text{Br} & \text{Br} & \text{Br} & \text{Br} \\
\text{I} & \text{II} & \text{III} & \text{IV}
\end{array}
$$

I and II are a pair of enantiomers as are III and IV. Diastereomer pairs are I and III, I and IV, II and III, and II and IV. There are no meso forms.

8. (2) 2,4-dibromopentane has a plane of symmetry and therefore exists as a meso form. (1) 2-bromobutane does not exist as a meso form.

CHAPTER TWENTY-SEVEN
Carbohydrates

SELF-EVALUATION SECTION

1. For each of the following carbohydrate molecules, pick out all of the correct descriptive terms from the list that apply to the particular molecule.

   triose
   tetrose
   pentose
   hexose
   number of asymmetric carbon atoms—
      e.g., 1, 2, 3, 4, or 5 atoms
   number of possible optical isomers—
      e.g., 2, 4, 8, 16 or 32 isomers

   reducing sugar
   nonreducing sugar
   D configuration
   L configuration
   aldose
   ketose

   Write your answers here.

   (1)   H   O
          \\ //
           C
           |
       H—C—OH
           |
          CH$_2$OH

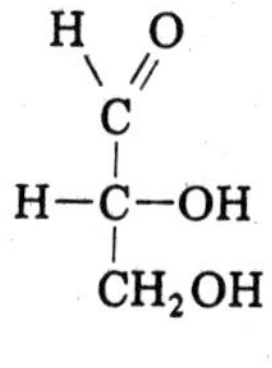

   (2)   CH$_2$OH
             |
         H—C—OH
             |
             C=O
             |
         H—C—OH
             |
            CH$_2$OH

271

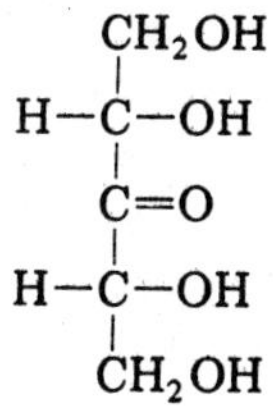

Write your answers here.

(3)
```
    H   O
     \ //
      C
      |
   H-C-OH
      |
  HO-C-H
      |
   H-C-OH
      |
   H-C-OH
      |
    CH₂OH
```

(4)
```
    CH₂OH
      |
      C=O
      |
   HO-C-H
      |
   H-C-H
      |
   H-C-OH
      |
    CH₂OH
```

(5)
```
    H   O
     \ //
      C
      |
   H-C-OH
      |
   H-C-OH
      |
  HO-C-H
      |
    CH₂OH
```

(6)
```
    H   O
     \ //
      C
      |
   H-C-OH
      |
   H-C-OH
      |
   H-C-OH
      |
    CH₂OH
```

(7)
```
    CH₂OH
      |
      C=O
      |
  HO-C-H
      |
    CH₂OH
```

(8)
```
    H   O
     \ //
      C
      |
   H-C-OH
      |
  HO-C-H
      |
  HO-C-H
      |
     CH₂OH
```

2. Fill in the blank space or circle the appropriate response.

Carbohydrates, one of the three principal classes of foods, contain only three elements: (1) _____, (2) _____, and (3) _____. The name "carbohydrate" is derived from the French *hydrates de carbone* because the empirical formula in many cases is (4) $C \cdot HO/(C \cdot H_2O)_n$. Carbohydrates are actually aldehydes or ketones that contain many (5) hydroxy/aromatic groups. The simplest carbohydrates are glyceraldehyde and dihydroxy acetone.

A classification system for carbohydrates is based on the number of units linked together. If there is only one unit, the carbohydrate is termed a (6) _____. Two units linked together are a (7) _____, such as sucrose, while long chain carbohydrates, such as glycogen and cellulose, are (8) _____. The most common monosaccharide, (9) _____, has (10) four/six asymmetric carbon atoms. Thus, the number of isomers related to glucose is (11) _____. All of these have been synthesized, but only D-glucose and D-galactose are of biological importance. Another common hexose, fructose, is a(n) (12) aldose/ketose, and when linked to glucose, it forms the disaccharide (13) _____. Other disaccharides are (14) _____ and lactose, which is known also as (15) _____. All monosaccharides and most disaccharides are reducing sugars capable of reducing silver ions and copper ions. The only common sugar that is not a reducing sugar is the disaccharide (16) _____.

3. Draw the possible isomers of a D-aldose that contains four carbons (tetrose): $C_4H_8O_4$ is the formula. Use open chain structural formulas.

Write your answers here.

4. Determine the oxidation states of each carbon atom in the first four carbohydrate molecules of Question 1. Number the uppermost carbon atom in each molecule as 1, the second as 2 and so on. Which carbon atom in each molecule may provide the most energy for a biological system?
Write your answers here:

5. Examine the carbohydrate formulas below and then answer the questions.

a.

b.
```
      H   OH
       \ /
        C
        |
     H—C—OH
        |
    HO—C—H     O
        |
     H—C—OH
        |
     H—C
        |
       CH₂OH
```

c. (pyranose ring with CH₂OH, OH groups)

d. (furanose ring with CH₂OH, OH, H)

e.
```
   H—C=O
     |
   H—C—H
     |
   H—C—OH
     |
   H—C—OH
     |
    CH₂OH
```

(1) Identify the pyranose and furanose ring structures.
(2) Identify the Fischer projection, "tree," and Haworth formulas.
(3) Identify the hemiacetal forms.
(4) How many carbons are there in the "tree" formula carbohydrate?
(5) Identify (a) and (e) as either D and/or L.

6. Match the disaccharides on the left with their two constituent monosaccharides from the list on the right.

    Disaccharides   **Monosaccharides**

    (1) sucrose   glucose
    (2) lactose   fructose
    (3) maltose   galactose

7. Match a possible source from the right with the sugar listed on the left.

    Sugar   **Source**

    (1) glucose   fruit juice
    (2) galactose   sugar beets
    (3) fructose   starch
    (4) sucrose   sucrose
    (5) lactose   lactose
    (6) maltose   pectin
                                         honey
                                         milk
                                         sugar cane
                                         sprouting grain

8. Match the polysaccharide with the correct descriptive statement.

    (1) starch   a. glucose units, α-1,4 linkage, highly branched
    (2) cellulose   b. fructose units, 1,2-linkage
    (3) glycogen   c. glucose units, α-1,4 linkage, moderately branched
    (4) inulin   d. glucose units, β-1,4 linkage

9. Fill in the blank space or circle the appropriate response.

The hemiacetal structure of carbohydrates involves an alcohol linkage and a(n)

(1) _____ linkage on the same carbon atom. Acetal structures

have (2) _____ ether linkages to the same carbon. The hemiacetal or cylic form is unstable and hydrolizes easily to the open form, which can reform to the other diastereomer. The symbols used for the two forms are

alpha (α) and beta (3) _____. Each of the diastereomers rotates polarized light and in solution there is an equilibrium between the forms. Diastereomers that differ in spatial arrangement only on C-1 carbon are called

(4) _____. The switching back and forth from α to β forms is

called (5) _____ and occurs in hemiacetals but not (6) _____
_____. Glycosides such as cellulose and starch are only ether linkages and are therefore (7) hemiacetal/acetal structures and (8) do/do not exhibit mutarotation. Lactose and maltose undergo mutarotation, which means at least one of their monosaccharides has a (9) _____ ring structure. Sucrose does not undergo mutarotation, which means only (10) _____ _____ linkages.

10. Complete the following reactions.

    (1) Benedict's test. What is visual evidence of a positive test?

    $$R-\underset{H}{\overset{H}{C}}=O + 2Cu^{2+} \text{ (blue)} \longrightarrow \underline{\hspace{2cm}} + \underline{\hspace{2cm}} + 3H_2O$$

    (2) Oxidation. For same tetrose, give products formed under two different oxidizing conditions.

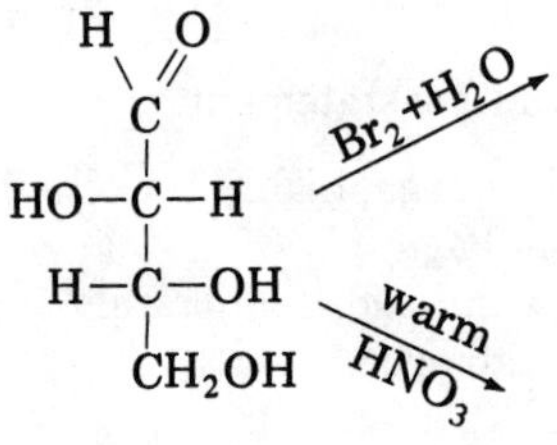

    $$\begin{array}{c} H\diagdown \;O \\ C \\ | \\ HO-C-H \\ | \\ H-C-OH \\ | \\ CH_2OH \end{array} \quad \xrightarrow[\text{warm } HNO_3]{Br_2 + H_2O}$$

    (3) Reduction of D-glucose with H_2/Pt

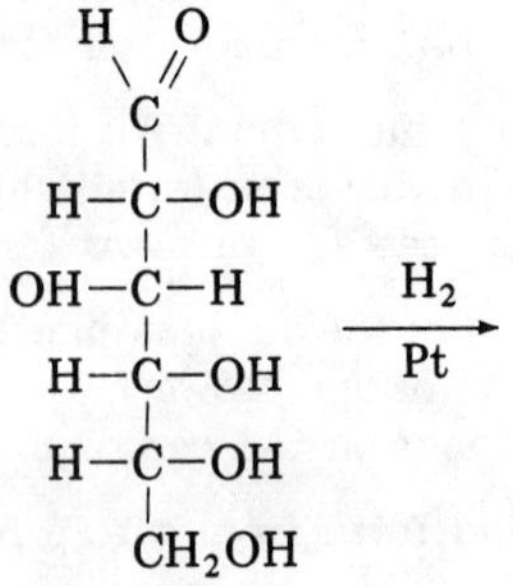

    $$\begin{array}{c} H\diagdown \;O \\ C \\ | \\ H-C-OH \\ | \\ OH-C-H \\ | \\ H-C-OH \\ | \\ H-C-OH \\ | \\ CH_2OH \end{array} \quad \xrightarrow{\dfrac{H_2}{Pt}}$$

11. Examine the "tree" formulas shown for several pentoses. Pick out the formulas that will give identical ozazones following reaction with phenyl hydrazine.

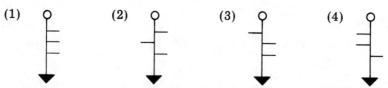

12. Using Table 27.1, indicate which functional group of the pairs given is higher in relative potential chemical energy.

   (1) CH$_3$ group versus primary alcohol
   (2) aldehyde versus methylene carbon
   (3) secondary alcohol versus ketone

RECAP SECTION

Chapter 27 has presented the structural information about one of three important classes of organic compounds utilized as foodstuffs—carbohydrates, fats, and proteins. We have learned the differences between monosaccharides, disaccharides, and polysaccharides, and the extensive number of monosaccharide isomers to be found because of asymmetric carbon atoms. The biochemical utilization of carbohydrates is the topic of Chapter 33 and we will apply what we've learned in the present chapter to a discussion of human digestion and metabolic processes.

ANSWERS TO QUESTIONS

1. (1) triose, 1 asymmetric carbon atom, 2 isomers, reducing sugar, D configuration, aldose.
   (2) pentose, 2 asymmetric carbon atoms, 4 isomers, nonreducing sugar, D configuration, ketose.
   (3) hexose, 4 asymmetric carbon atoms, 16 isomers, reducing sugar, D configuration, aldose.
   (4) hexose, 2 asymmetric carbon atoms, 4 isomers, reducing sugar, D configuration, ketose.
   (5) pentose, 3 asymmetric carbon atoms, 8 isomers, reducing sugar, L configuration, aldose.
   (6) pentose, 3 asymmetric carbon atoms, 8 isomers, reducing sugar, D configuration, aldose.
   (7) tetrose, 1 asymmetric carbon atom, 2 isomers, reducing sugar, L configuration, ketose.
   (8) pentose, 3 asymmetric carbon atoms, 8 isomers, reducing sugar, L configuration, aldose.

2. (1) carbon   (2) oxygen   (3) hydrogen (in any order)
   (4) (C·H$_2$O)$_n$ (5) hydroxy   (6) monosaccharide
   (7) disaccharide   (8) polysaccharides   (9) glucose   (10) four
   (11) 16   (12) ketose   (13) sucrose   (14) maltose
   (15) milk sugar   (16) sucrose

3. Since we are dealing with a tetrose, which is an aldose, we will have two asymmetric carbon atoms at carbons number 2 and 3. We have a D sugar, which means the hydroxy group on carbon number 3 is to the right.

$$\begin{array}{c} HO \\ \diagdown\diagup \\ C \\ | \\ -C- \\ | \\ H-C-OH \\ | \\ CH_2OH \end{array}$$

If there are two asymmetric carbon atoms, then there are 2^2 or four isomers, two pairs of enantiomers. The two D isomers are, therefore,

$$\begin{array}{c} HO \\ \diagdown\diagup \\ C \\ | \\ H-C-OH \\ | \\ H-C-OH \\ | \\ CH_2OH \end{array} \quad \text{and} \quad \begin{array}{c} HO \\ \diagdown\diagup \\ C \\ | \\ HO-C-H \\ | \\ H-C-OH \\ | \\ CH_2OH \end{array}$$

4. (1) $C_1 = +1$, $C_2 = 0$, $C_3 = -1$.
    C_3 is most reduced and should provide most energy.
   (2) $C_1 = -1$, $C_2 = 0$, $C_3 = +2$, $C_4 = 0$, $C_5 = -1$.
    C_1 and C_5 should provide the most energy.
   (3) $C_1 = +1$, $C_2 = 0$, $C_3 = 0$, $C_4 = 0$, $C_5 = 0$, $C_6 = -1$.
    C_6 should provide the most energy.
   (4) $C_1 = -1$, $C_2 = +2$, $C_3 = 0$, $C_4 = -2$, $C_5 = 0$, $C_6 = -1$.
    C_4 should provide the most energy.

5. (1) b and c are pyranose, d is furanose.
   (2) Fischer projections are b and e, tree formula is a, and Haworth formulas are c and d.
   (3) Hemiacetal forms are b, c, d.
   (4) Tree formula has six carbons, a hexose.
   (5) a is "D" and e is "D."

6. (1) sucrose—glucose and fructose
   (2) lactose—glucose and galactose
   (3) maltose—two glucose units

7. (1) glucose—starch, sucrose, lactose, and maltose
   (2) galactose—lactose, pectin
   (3) fructose—honey, sucrose, fruit juice
   (4) sucrose—sugar cane, sugar beets
   (5) lactose—milk
   (6) maltose—sprouting grain

8. (1) starch—c (3) glycogen—a
   (2) cellulose—d (4) inulin—b

9. (1) ether (2) two (3) β (4) anomers
   (5) mutarotation (6) acetals (7) acetal (8) do not
   (9) hemiacetal (10) acetal or ether

10. (1) Products would be R—C(=O)(O—) and Cu_2O

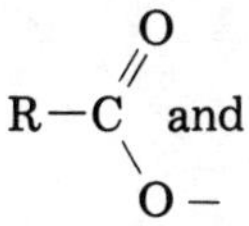

   The Cu_2O will be a brick-red precipitate.

   (2) Product from $Br_2 + H_2O$ oxidation would be

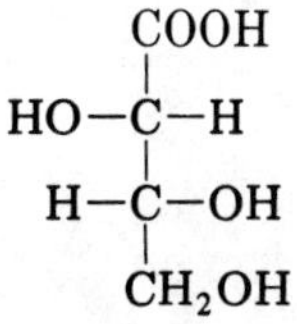

   COOH
   |
   HO—C—H
   |
   H—C—OH
   |
   CH_2OH

   Product from warm HNO_3 oxidation would be

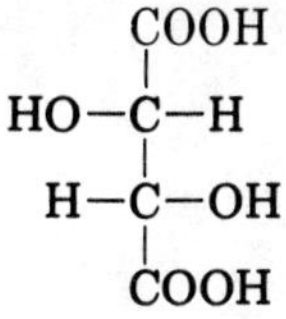

   COOH
   |
   HO—C—H
   |
   H—C—OH
   |
   COOH

   (3) Reduction of the glucose product is

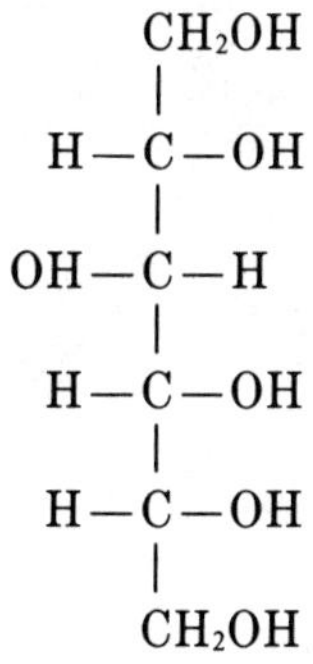

   CH_2OH
   |
   H—C—OH
   |
   OH—C—H
   |
   H—C—OH
   |
   H—C—OH
   |
   CH_2OH

11. (1) and (3) would give identical ozazones
    (2) and (4) would also give identical ozazones

12. (1) CH_3 group (2) methylene carbon (3) secondary alcohol

CHAPTER TWENTY-EIGHT
Lipids

SELF-EVALUATION SECTION

1. Fill in the blank space or circle the appropriate response.

Lipids are a group of organic compounds classified on the basis of solubitily in fat solvents such as (1) _____. The simple lipids are the fats, oils, and waxes. Fats and oils are fatty acid esters of the trihydroxy alcohol (2) _____. The long-chain R groups of a fat are usually (3) <u>the same/different</u>. The main physical difference between fats and oils is that fats are (4) _____ at room temperature, while oils are (5) _____. In addition, fats usually come from (6) <u>vegetable/animal</u> sources, while oils are from (7) _____ sources. The difference in physical properties in part can be attributed to the amount of unsaturation in the fatty acid groups. Oils generally have a (8) <u>greater/lesser</u> amount of unsaturation as compared with fats. Oils are hardened by the process of hydrogenation to make them solid. During the reaction, hydrogen is (9) <u>added to/removed from</u> the carbon-carbon double bonds. When compared with carbohydrates and proteins, fats provide a (10) <u>greater/lesser</u> amount of caloric value.

Waxes are (11) <u>complex/simple</u> esters of (12) <u>long-/short-</u>chain fatty acids and alcohols. They are usually of (13) <u>animal/plant</u> origin and may serve protective functions such as helping reduce water loss from the leaves. Phospholipids, glycolipids, and steroids are examples of (14) <u>simple/complex</u> lipids that serve many regulatory biological functions as vitamins and hormones.

2. Identify the hydrophobic and hydrophilic portions of the following phospholipid.

$$\begin{array}{c} \text{CH}_2-\text{O}-\overset{\overset{\displaystyle O}{\|}}{\text{C}}-\text{R} \\ \overset{\overset{\displaystyle O}{\|}}{\text{R}-\text{C}}-\text{O}-\text{CH} \\ \text{CH}_2-\text{O}-\overset{\overset{\displaystyle O}{\|}}{\underset{\underset{\displaystyle \text{O}^-}{|}}{\text{P}}}-\text{O}^- \end{array}$$

3. How does aspirin affect the conversion of arachidonic acid to prostaglandins?

4. Draw the structure common to all steroids.

5. Determine the oxidation number of the central carbon in the following species and indicate which has the greater potential chemical energy and why.

$$\text{carbohydrate fragment} \qquad \begin{array}{c} \text{OH OH OH} \\ | | | \\ \text{---C}-\text{C}-\text{C---} \\ | | | \\ \text{H} \text{H} \text{H} \end{array}$$

$$\text{fat fragment} \qquad \begin{array}{c} \text{H} \text{H} \text{H} \\ | | | \\ \text{---C}-\text{C}-\text{C---} \\ | | | \\ \text{H} \text{H} \text{H} \end{array}$$

6. Match the phrase of key words from the right hand list with the terms on the left:
   - (1) atherosclerosis
   - (2) essential fatty acids
   - (3) phospholipid
   - (4) cholesterol
   - (5) fat
   - (6) cephalin
   - (7) sphingomyelin
   - (8) glycolipids
   - (9) cerebrosides
   - (10) anesthetic
   - (11) ganglioside
   - (12) sphingosine
   - (13) membranes

   - a. linoleic, arachidonic, linolenic
   - b. myelin sheath
   - c. converted to ACTH, estrogen and testosterone
   - d. deposition of cholesterol as plaque
   - e. thin, semi-permeable cellular barriers
   - f. lecithins
   - g. major body reserve of potential energy
   - h. essential in blood clotting
   - i. a glycolipid containing complex oligosaccharides as their carbohydrate
   - j. cell membrane of brain tissue
   - k. sphingolipids that contain carbohydrates
   - l. unsaturated amino alcohol
   - m. alters membrane fluidity

7. Below is a diagram of a typical membrane. Identify each section as being either hydrophobic or hydrophilic.

   _____ a
   ///////////////////////// b
   _____ c

8. Fill in the blank space or circle the appropriate response.

   The chemical composition of membranes, a lipid (1) <u>bilayer/monolayer</u>, gives rise to questions about how hydrophilic molecules are transported across the membrane from one side to the other. It is now known that specific (2) <u>alcohols/proteins</u> occur within the bilayer and allow for the movement of hydrophilic molecules across the lipid bilayer. If the protein helps the transport without using energy, the process is termed (3) _____ while the movement of molecules from areas of low concentration to high concentration is called (4) _____ and requires energy.

   Nerve cells provide good examples of membrane function. An active transport system, which (5) <u>uses/produces</u> energy, concentrates K^+ ions inside the cell while expelling Na^+ ions. When a neuron transmits a signal, some K^+ ions flow out rapidly and some Na^+ flows in without using energy. This is an example of (6) _____. The rapid ion movement is an electrical nerve impulse.

   The myelin sheath acts as an (7) <u>conductor/insulator</u> around the neuron and when the myelin sheath is damaged or destroyed, nerve transmission is poor resulting in a serious crippling disease, (8) _____.

RECAP SECTION

Chapter 28 presents information about the class of biochemical compounds called lipids. Lipids serve important structural functions in membranes and as potential energy sources for metabolism. Several detailed examples were presented to give you an appreciation of lipid compounds importance for cells and organisms.

ANSWERS TO QUESTIONS

1. (1) ether, benzene, or carbon tetrachloride
   (2) glycerol   (3) different   (4) solid
   (5) liquid   (6) animal   (7) vegetable
   (8) greater   (9) added to   (10) greater
   (11) simple   (12) long   (13) plant
   (14) complex

2. The $-O-\overset{\overset{O}{\|}}{C}-R$ portion is the hydrophobic part

   The $-O-\overset{\overset{O}{\|}}{\underset{\underset{O^-}{|}}{P}}-O^-$ portion is the hydrophilic part

3. Aspirin prevents the enzymatic oxidation of arachidonic acid to prostaglandins.

4.

5. Oxidation number of carbon in the carbohydrate is zero (0) and in the fat the number is −2. Therefore, carbon in a fat is more reduced than in carbohydrate and contains more potential energy. Fats are better sources of energy than carbohydrates because of more reduced carbon.

6. (1) d   (2) a   (3) f   (4) c   (5) g   (6) h
   (7) b   (8) k   (9) j   (10) m   (11) i   (12) l   (13) e

7. membrane   (a) hydrophilic   (b) hydrophobic   (c) hydrophilic

8. (1) bilayer   (2) proteins   (3) facilitated diffusion   (4) active transport
   (5) uses   (6) facilitated diffusion   (7) insulator   (8) MS (multiple sclerosis)

CHAPTER TWENTY-NINE
Amino Acids, Polypeptides, and Proteins

SELF-EVALUATION SECTION

1. From the list below, circle the foods high in protein content.

bread	melon	spaghetti	apples
fish	chicken	cheese	potatoes
carrots	nuts	eggs	beans

2. Match the term in Column A with the description in Column B.

Column A		Column B
(1) Primary structure	_____	a. Equal number of amino and carboxyl groups.
(2) Basic amino acid	_____	b. The type and sequence of amino acids in a protein.
(3) Neutral amino acid	_____	c. The higher order of structure found in complex proteins.
(4) Tertiary structure	_____	d. More carboxyl groups than amino groups.
(5) Acidic amino acid	_____	e. More amino groups than carboxyl groups.
(6) Quaternary structure	_____	f. The characteristic shape or conformation of a protein.
(7) Secondary structure	_____	g. The pleated or helical structure of a protein.

3. Give the product ion when the valine zwitterion reacts with (1) acid and (2) base.

285

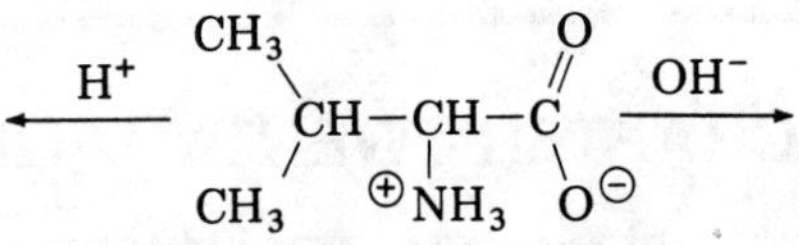

(1)

$$\xleftarrow{H^+} \begin{array}{c} CH_3 \\ \backslash \\ CH-CH-C \\ / | \backslash \\ CH_3 {}^{\oplus}NH_3 O^{\ominus} \end{array} \begin{array}{c} O \\ \| \end{array} \xrightarrow{OH^-}$$

(2)

4. Fill in the blank space with either the name of the amino acid or its abbreviation.

Name	Abbreviation
(1) Glutamic acid	
(2)	val
(3)	ser
(4) Leucine	
(5) Cysteine	
(6) Histidine	
(7)	arg
(8)	pro
(9) Methionine	
(10) Phenylalanine	

5. Fill in the blank space or circle the appropriate response.

Proteins, which are polymers composed of (1) _____, serve two essential functions: that of (2) <u>a source of energy/a structural role</u> and (3) <u>as enzymes/as activators</u>. When proteins are hydrolyzed into their component amino acids, scientists find approximately (4) <u>30/20</u> usual amino acids with (5) <u>two/three</u> identical functional groups per amino acid. These two groups are (6) _____ and (7) _____. On a list of common amino acids there are eight which are classified as essential for (8) <u>beef/humans</u>. This means that these amino acids cannot be (9) _____ by an individual and must be supplied in the (10) _____. The physical and chemical properties of proteins are determined largely by the (11) _____ of the amino acids. In addition to serving a variety of structural roles in an organism, proteins are also found to be biological catalysts or (12) _____. These specialized molecules reduce the (13) <u>concentration/ energy of activation</u> of biological reactions and allow the reactions to proceed at body temperatures and neutral pH. In addition to proteins, another class of biologically active molecules has been recently discovered to act as enzymes in certain cases. These molecules are (14) _____. The molecules upon which enzymes act are called (15) _____.

6. Write the dipeptide structure formed when glycine is joined to alanine. Point out the peptide linkage. Name the dipeptide.

7. Name the following pentapeptides.

   (1) phe-thr-pro-leu-gly

   (2) his-gly-ala-tyr-val

8. An octapeptide was known to contain two proline residues and two leucine residues. The following peptides were obtained after partial hydrolysis. Determine the sequence.

       ala-pro-his, pro-leu, ser-leu, his-pro, pro-leu-val-ser

9. Match the term on the right with a phrase from the list on the left.

   (1) ninhydrin test   a. removal of groups from substrate
   (2) Sanger   b. changes in protein properties without
   (3) peptide configuration   hydrolysis
   (4) elements found in proteins   c. catalyze only one reaction
   (5) cysteine   d. C, H, O, S, N
   (6) Biuret test   e. concentrated nitric acid
   (7) chromatographic separation   f. α-helix and β-pleated sheet
        techniques   g. disulfide bonding
   (8) absolute specificity   h. violet color with copper sulfate
   (9) hydrolases   i. amino acid sequence in insulin
 (10) xanthroproteic reaction   j. thin layer, paper, column
 (11) denaturation   k. catalyze hydrolysis of esters, etc.
 (12) lyases   l. blue color except with proline and
                                                   hydroxyproline.

10. Quaternary structure of hemoglobin is important in the (1) <u>oxygen/hydrogen</u> transport system of the blood. When the hemoglobin, composed of (2) _____ subunits, binds one molecule of oxygen to itself, its conformation changes to (3) <u>discourage/facilitate</u> the binding of (4) _____ additional oxygen molecules. As hemoglobin moves from the lungs to cells needing oxygen, one

oxygen is removed. The protein (5) _____ changes again and the (6) _____ remaining oxygen molecules are then more easily removed. The presence of a (7) _____ structure allows the binding/removal of one oxygen molecule to control the binding/removal of three other oxygen molecules.

CHALLENGE PROBLEM

11. As much as possible without looking at the text, examine the combined Michaelis-Menten plots for two enzyme-catalyzed reactions under identical conditions and label the various axes and points on the graph and experimental curves as called for.

    Then compare (a) the catalytic efficiency of the two enzymes and (b) their respective effect substrate concentration ranges.

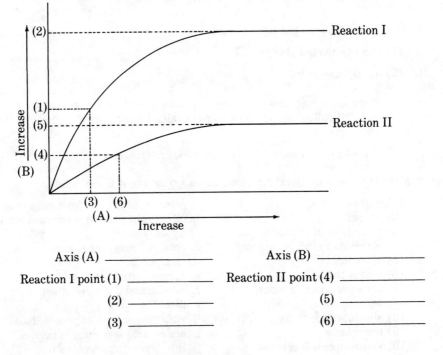

Axis (A) _____   Axis (B) _____
Reaction I point (1) _____   Reaction II point (4) _____
             (2) _____   (5) _____
             (3) _____   (6) _____

RECAP SECTION

Chapter 29 has given us insights into one of the many fascinating areas of biochemical research. Much of the work has been of a very creative nature and has allowed scientists to begin to see how protein structure, enzyme reactivity, and genetics are interrelated. These molecules are very complex in their structure, yet there is an orderliness about their synthesis that is fascinating. Chapter 29 takes us further in the study of the genetic molecules themselves and how their information is passed on to new organisms.

ANSWERS TO QUESTIONS

1. fish, chicken, nuts, cheese, eggs, beans

2. (1) b   (2) e   (3) a   (4) f
   (5) d   (6) c   (7) g

3. (1)
```
   CH₃          O
     \         ∥
      CH—CH—C
     /    |    \
   CH₃   ⊕NH₃   OH
```
   (2)
```
   CH₃          O
     \         ∥
      CH—CH—C
     /    |    \
   CH₃   NH₂    O⊖
```

4. (1) glu   (2) Valine   (3) Serine   (4) leu
   (5) cys   (6) his   (7) Arginine   (8) Proline
   (9) met   (10) phe

5. (1) amino acids   (2) a structural role   (3) as enzymes
   (4) 20   (5) two   (6) amino
   (7) carboxyl   (8) humans   (9) synthesized
   (10) diet or food   (11) side chains   (12) enzymes
   (13) energy of activation   (14) ribonucleic acids   (15) substrates

6.
```
                 peptide linkage
           O    /       O
           ∥   /        ∥
   CH₂—C—NH—CH—C          glycyl alanine
    |         |   \
   NH₂       CH₃   OH
```

7. (1) phenylalanylthreonylprolylleucylglycine
   (2) histidylglycylalanyltyrosylvaline

8. The octapeptide would be

   ala-pro-his-pro-leu-val-ser-leu

9. (1) l   (2) i   (3) f   (4) d   (5) g
   (6) h   (7) j   (8) c   (9) k   (10) e
   (11) b   (12) a

10. (1) oxygen   (2) four   (3) facilitate   (4) three
    (5) conformation or shape   (6) three   (7) quaternary

11. Axis (A) is the substrate concentration
    Axis (B) is the rate of reaction or change in concentration per minute.

    Reaction I, point (1) is 1/2 V_{max} (rate)
                   (2) is V_{max} (rate)
                   (3) is K_m, Michaelis constant

Reaction II, point (4) is $1/2\ V_{max}$ (rate)
              (5) is V_{max} (rate)
              (6) is K_m, Michaelis constant

Reaction I has a higher V_{max} than Reaction II. Therefore I is catalyzed by the more efficient enzyme. Since K_m is a measure of effective substrate concentration range, a lower K_m value means an enzyme works well at lower substrate concentrations. Thus the enzyme in Reaction I is more effective than the enzyme in Reaction II.

CHAPTER THIRTY
Nucleic Acids and Heredity

SELF-EVALUATION SECTION

1. Fill in the blank space or circle the correct response.

The great variety of proteins found in any organism does not arise by chance. The substances that direct the course of protein synthesis are known as (1) _____. The two major types are known by abbreviations, (2) _____ and (3) _____. There are several major differences between these types of nucleic acids. RNA contains the sugar (4) _____, while DNA contains (5) _____. DNA is a double-stranded (6) _____, while RNA is only (7) _____ stranded. In addition, one of the pyrimidine bases is different. DNA contains the base (8) _____ and RNA contains (9) _____. The term nucleoside refers to either the purine or pyrimidine base joined with a (10) _____ molecule. There are (11) 10/15 such nucleosides, each with a specific abbreviation. Thus A stands for (12) _____, dG stands for (13) _____, T stands for (14) _____, and dU stands for (15) _____. In nucleic acids, the actual building blocks are phosphate esters of nucleosides. These molecules are called (16) _____. Again, there are specific abbreviations. AMP stands for (17) _____, ADP for (18) _____, and ATP for (19) _____. These last two nucleotides are extremely important biological compounds as easily obtainable chemical energy resides in their hydrolyzable (20) _____ bonds. Thus, when ATP is hydrolyzed to ADP and inorganic phosphate via an enzyme catalyzed reaction, a considerable amount of energy is released.

In the double-stranded helix of DNA, it has been found that (21) <u>complementary/dissimilar</u> pairing occurs between purine and pryimidine bases. dA pairs with (22) _____ and dG pairs with (23) _____. The genetic information contained in the cell's genes is found to be the exact sequence of nucleotides along the DNA strands. When the DNA replicates itself during normal cell division called (24) _____, the two daughter cells have the same double-stranded DNA sequences as the original cell. During sex cell production, called (25) _____, only half the genes or DNA sequences are present in the new cells. When fertilization takes place the zygote has a normal complement of genes, half from the mother and half from the father. Each gene, which directs the synthesis of one protein, contains what is termed the genetic code. Each codon, for a specific amino acid, is composed of (26) _____ nucleotides in sequence. The intermediate compound, which carries the information from the nucleus (site of the DNA) to the ribosomes (site of protein synthesis), is a type of ribonucleic acid called (27) _____. The ribonucleic acid, which brings various amino acids to the ribosomes, is called (28) _____ and has the shape of a (29) _____. Each loop of a particular t-RNA has a purpose.

If, for some reason, the DNA structure is changed or a breakdown occurs in the protein synthesis process, a (30) _____ may occur. Most severe cases result in an organism's death, but sometimes the organism survives. For example, sickle cell anemia results from (31) <u>two/one</u> change(s) in an amino acid in one chain of hemoglobin.

2. Match the term on the left with a phrase or term on the right.

(1) guanine
(2) cytosine
(3) transfer RNA
(4) uracil
(5) messenger RNA
(6) ribosomal RNA
(7) adenine
(8) thymine
(9) genome
(10) oncogenes
(11) AUG and GUG
(12) GTP

a. purine base
b. causes cells to become cancerous
c. pyrimidine base
d. carries genetic information from DNA to ribosomes
e. condons for the start of protein synthesis
f. brings amino acids to the ribosomes for incorporation into protein
g. nucleotide that is primary source of energy for protein synthesis
h. part of structure at which protein synthesis occurs
i. sum of all hereditary material contained in a cell

CHALLENGE PROBLEM

3. Try to fill in the boxes or spaces with the correct terms without using the text.

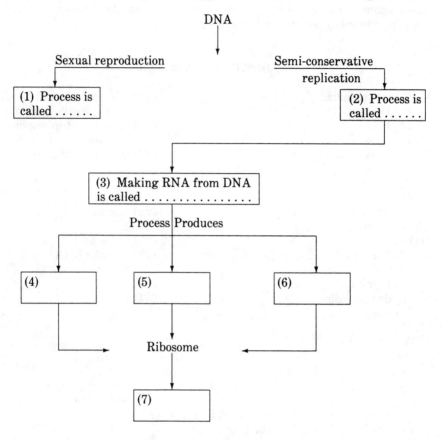

The ribosomes are the site of (7) _____ synthesis, a process called (8) _____. A primary function of the tRNA is to bring (9) _____ to the ribosomes for incorporation into proteins. The messenger RNA contains the information or (10) _____ in the form of three (11) _____. Some amino acids have only one codon each but for other amino acids the code is (12) _____.

RECAP SECTION

Chapter 30 is an excellent introduction to a very complex and fascinating subject. The science of genetics and the biochemical basis for heredity and the genetic code are current areas of research today. Much has been learned in the last 25 years about the process of protein synthesis and several genetic diseases have been explained and are now treatable. The future promises other problem areas now that researchers are able to alter the genetic material and induce various types of mutations. Cloning of organisms is another concern. An informed public is of great importance in the area of DNA research.

ANSWERS TO QUESTIONS

1. (1) nucleic acids
   (2) DNA
   (3) RNA
   (4) ribose
   (5) 2-deoxy-D-ribose
   (6) helix
   (7) single
   (8) thymine
   (9) uracil
   (10) sugar
   (11) 10
   (12) adenosine
   (13) deoxyguanosine
   (14) thymidine
   (15) deoxyuridine
   (16) nucleotides
   (17) adenosine monophosphate
   (18) adenosine diphosphate
   (19) adenosine triphosphate
   (20) phosphate
   (21) complementary
   (22) dT
   (23) dC
   (24) mitosis
   (25) meiosis
   (26) three
   (27) messenger RNA
   (28) transfer RNA
   (29) cloverleaf
   (30) mutation
   (31) one

2. (1) a (2) c (3) f (4) c (5) d (6) h
   (7) a (8) c (9) i (10) b (11) e (12) g

3. (1) meiosis   (2) mitosis   (3) transcription
   (4) mRNA   (5) rRNA   (6) tRNA
                 (4), (5), (6) in any order
   (7) protein   (8) translation   (9) amino acids
   (10) code   (11) nucleotides   (12) redundant

CHAPTER THIRTY-ONE
Nutrition and Digestion

SELF-EVALUATION SECTION

1. Fill in the missing places in the following table.

Digestive Fluid Source	Fluid Name	Principal Enzymes
Salivary glands	(1) _____	(2) _____
(3) _____	(4) _____	Pepsin, rennin, gastric lipase
(5) _____	Pancreatic juice	Trypsin, chymotrypsin
(6) _____	(7) _____	Bile salts
Glands in duodenum	(8) _____	Finishing enzymes

2. Match the digestive fluid with the type of food digestion catalyzed or facilitated by the particular fluid.

   (1) gastric
   (2) bile juice
   (3) pancreatic juice
   (4) saliva

   a. digestion of all three kinds of food
   b. starch hydrolysis, lubrication
   c. protein digestion
   d. fat digestion

3. List the six nutrient groups.

4. List the six food groups.

5. Name at least 3 of the major elements and 3 of the trace elements.

6. Match the phrases or descriptions on the right with the terms on the left.

   (1) water
   (2) vitamins
   (3) marasmus
   (4) RDA
   (5) obesity
   (6) Kwashiorkor
   (7) cellulose
   (8) complete protein
   (9) diet

   a. recommended dietary allowance
   b. supplies all essential amino acids
   c. excess calories
   d. calorie deficiency
   e. food and drink we consume
   f. dietary fiber
   g. components of enzyme systems
   h. body fluids, reaction solution
   i. protein deficiency

7. To lose one pound of body fat requires an energy expenditure of 3500 kcal greater than the energy intake.
   If a person could realistically reduce their energy allowance by 500 kcal per day, how many days would be required to lose 30 pounds?

   Do your calculations here.

8. List the four purposes of food additives and give an example of an additive for each purpose.

RECAP SECTION

Chapter 31 discusses the biochemical aspects of proper nutrition and some of the consequences of improper diet. The relationships between various nutrients in a balanced diet are described as is the process of digestion. These topics are of general interest to us all and of special importance to persons in the health fields.

ANSWERS TO QUESTIONS

1. (1) saliva   (2) amylase   (3) stomach glands   (4) gastric juice
   (5) pancreas   (6) liver   (7) bile   (8) intestinal juice

2. (1) c   (2) d   (3) a   (4) b

3. carbohydrates, lipids, proteins, water, minerals, vitamins.

4. milk products, vegetables, fruits, cereal products, meats, fatty foods.

5. Na, K, Ca, Mg, Cl, P are the major elements.
   F, Si, Cr, Fe, Co, Ni, Cu, Zn are examples of the trace elements.

6. (1) h   (2) g   (3) d   (4) a   (5) c   (6) i   (7) f   (8) b   (9) e

7. 210 days or 30 weeks

$$\frac{(30 \text{ pounds}) (3500 \text{ kcal/pound})}{500 \text{ kcal/day}} = 210 \text{ day}$$

8. (1) enhance nutritional value—vitamins, minerals
   (2) preservative—sodium benzoate, calcium lactate, sorbic acid
   (3) anti-oxidant—BHA, BHT
   (4) improve appearance and flavor—pectin, propylene glycol

CHAPTER THIRTY-TWO
Bioenergetics

SELF-EVALUATION SECTION

1. Fill in the blanks or circle correct response.

    The sum of all chemical changes that occur in a living organism is given the name (1) _____. Further, these changes have two aspects. When a process involves simple substances being synthesized into complex substances, the process is termed (2) _____. The opposite process, involving the breaking down of complex materials into simpler substances, is termed (3) _____. The raw biochemical materials used as sources for the metabolic energy processes are classed as (4) _____, (5) _____ and (6) _____. These energy sources originally were derived from the energy of (7) foodstuffs/sunlight by a process in green plants called (8) _____. Energy is made available to the cell by means of (9) displacement/oxidation-reduction reactions which involve transfer of (10) _____. The organelle which is the site of most oxidation-reduction reactions is the (11) _____. The transport of electrons from one reaction site to another involves various coenzymes derived from B-complex vitamins. The three most common coenzymes in electron transport are (12) _____, (13) _____ and (14) _____. The released energy is stored in a phosphate anhydride bond in the molecule (15) _____.

2. Name two advantages to a cell for using ATP as a stored form of energy.

3. Identify the following phrases as applying to either substrate-level phosphorylation (SLP) or to oxidative phosphorylation (OP).

    (1) oxidation of reduced carbon to form high energy phosphate bonds _____.
    (2) direct use of oxidation-reduction reactions to form ATP _____.
    (3) site is mitochondria and involves electron transport system _____.
    (4) involves phosphorylated substrates _____.

4. Identify the following phrases as applying to either mitochondria or chloroplasts or both:

   (1) contains lipid
   (2) site of electron transport system
   (3) action of electron transport system releases energy
   (4) action of electron transport system traps energy
   (5) reduction of coenzymes with the input of energy
   (6) oxidation of coenzymes with the output of energy

5. List 4 different subcellular organelles.

6. Match the phrase or key words from the right-hand list with the terms on the left.

   (1) procaryote cell
   (2) eucaryote cell
   (3) liver
   (4) membrane-bound ribosomes
   (5) kidneys
   (6) mitochondria
   (7) chloroplasts
   (8) red blood cell
   (9) oxidative phosphorylation
   (10) photosynthesis

   a. component responsible for rapid oxygen transport
   b. site of photosynthesis
   c. production site of most cellular energy molecules
   d. produce proteins for secretion from the cell
   e. lacks distinct membrane-bound nucleus
   f. converts lactic acid to glucose
   g. membrane-bound organelles and nucleus
   h. transfer water from bloodstream to urine
   i. requires oxygen
   j. source of nearly all biochemical system energy

7. Fill in the blank space with an appropriate term.
   A good example of a coordinated cell function is the removal of carbon dioxide from muscle cells to the (1) _____ where it is exhaled. This process is tied to the release of oxygen from the transport molecule (2) _____ at a site of use, the muscle cells. During exertion, the muscle cells produce CO_2 and require O_2. Carbon dioxide reacts with amino groups on the (3) _____ molecule and releases hydrogen ions. The hydrogen ions promote the release of more oxygen from oxyhemoglobin to be used by the muscle cells. A second removal process for carbon dioxide also involves the production of hydrogen ions in the muscle cells. In the lungs, the hemoglobin is reoxygenated. Carbon dioxide is released by the reactions in the lungs and is exhaled.

   The liver and the kidneys also play important roles in maintaining the proper acid (hydrogen ion) level in the body. The liver replaces reduced carbon energy sources in the blood streams as muscle cells increase their activity. During hard work, muscle cells will function in an oxygen deficient environment temporarily and produce (4) _____ instead of carbon dioxide. The liver converts this product back to (5) _____ for recycling to the muscle cells. The kidneys, a prime regulator of water and ion balance, selectively transfer water and ions from the blood to the (6) _____ which is a slow process so that kidney function is important for the long term maintenance of correct body fluid pH values.

8. Match phrase in left-hand column with correct phrase in right-hand column:

Energy use at muscle	How gas balance restored in the lungs
(1) O_2 used	a. CO_2 exhaled
(2) CO_2 formed	b. more O_2 bound to Hb
(3) acidity increases	c. O_2 inhaled
(4) more O_2 released from HbO_2	d. acidity decreases

SUMMARY

Chapter 32 ties together a number of topics which have been discussed briefly earlier in the text. The interrelationship of cellular processes involving energy transformations in living organisms is a very important subject for a wide variety of students. The application of basic chemical principles to biological problems is a worthwhile and exciting endeavor.

ANSWERS TO QUESTIONS

1. (1) metabolism   (2) anabolism   (3) catabolism   (4) carbohydrates
   (5) lipids   (6) proteins   (7) sunlight   (8) photosynthesis
   (9) oxidation-reduction   (10) electrons   (11) mitochondria
   (12) NAD^+ (13) $NADP^+$ (14) FAD   (15) ATP

2. (1) Energy stored in ATP readily accessible through a simple hydrolysis reaction.
   (2) Energy from various types of reactions is funneled into one storage form, ATP. ATP is a common energy currency in a cell.

3. (1) SLP   (2) OP   (3) OP   (4) SLP

4. (1) mitochondria and chloroplast
   (2) mitochondria and chloroplast
   (3) mitochondria
   (4) chloroplast
   (5) chloroplast
   (6) mitochondria

5. Possibilities are: nucleus, ribosome, mitochondrion, chloroplasts, lysome, peroxisome, Golgi apparatus

6. (1) e   (2) g   (3) f   (4) d   (5) h   (6) c   (7) b   (8) a   (9) i   (10) j

7. (1) lungs (2) hemoglobin (3) hemoglobin (4) lactic acid (5) glucose (6) urine

8. (1) c   (2) a   (3) d   (4) b

CHAPTER THIRTY-THREE
Carbohydrate Metabolism

SELF-EVALUATION SECTION

1. Fill in the missing places in the following table.

Digestive Fluid Source	Fluid Name	Principal Enzymes
Salivary glands	(1) _____	(2) _____
(3) _____	(4) _____	Pepsin, rennin, gastric lipose
(5) _____	Pancreatic juice	Trypsin, chymotrypsin
(6) _____	(7) _____	Bile salts
Glands in duodenum	(8) _____	Finishing enzymes

2. Match the digestive fluid with the type of food digestion catalyzed or facilitated by the particular fluid.

   (1) gastric
   (2) bile juice
   (3) pancreatic juice
   (4) saliva

   a. digestion of all three kinds of food
   b. starch hydrolysis, lubrication
   c. protein digestion
   d. fat digestion

3. Fill in the blank space or circle the appropriate response.

   In the body, glucose can be oxidized to carbon dioxide and (1) _____ _____ with the trapping of the energy in the form of phosphate bond energy in the transfer molecule (2) _____. Or glucose can be stored as (3) _____ to function as a reserve.

   The anaerobic conversion of glucose to pyruvic acid is called the (4) _____ _____ pathway. The process is called anaerobic because no (5) _____ _____ is involved. Once pyruvic acid is formed it enters an aerobic cycle known as the (6) _____ cycle, which eventually extracts most of the chemical bond energy originally present in the glucose.

Whatever the fate of glucose and other carbohydrate molecules during metabolism, each of the reactions is catalyzed by specific enzymes, and in many cases controlled by regulatory molecules called (7) _____. These molecules are secreted by the ductless or (8) _____ glands and are synthesized by the body. The key molecule in the metabolic relationship between amino acids, fatty acids, and the citric acid cycle is (9) _____.

4. Match the terms on the left with one or more items from the list on the right.

   (1) Hyperglycemia
   (2) Hypoglycemia
   (3) Renal threshold
   (4) Insulin
   (5) Epinephrine

   a. reduces blood glucose
   b. below fasting level
   c. increases blood glucose
   d. glucose above fasting level
   e. increases rate of glycogen breakdown
   f. increases rate of glycogen formation
   g. glucose appears in urine
   h. 90-140 mg glucose per 100 ml of blood

5. The diagram below represents an overview of carbohydrate metabolism. From the list of terms given, identify A, D, C and D.

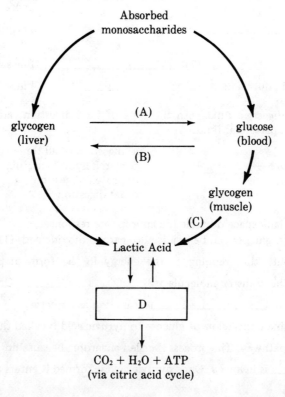

Terms to be used for diagram above and the diagram in question number 6 which follows:

(1) ketoglutaric acid
(2) glycolysis
(3) ATP
(4) acetyl-CoA
(5) glycogenesis
(6) glycine
(7) pyruvic acid
(8) amino acids
(9) glycogenolysis
(10) fatty acids
(11) NADH
(12) chlorophyll
(13) glucose
(14) CO_2

6. The diagram below represents a simplified illustration of the interrelationship of carbohydrates, fats and proteins in the body.

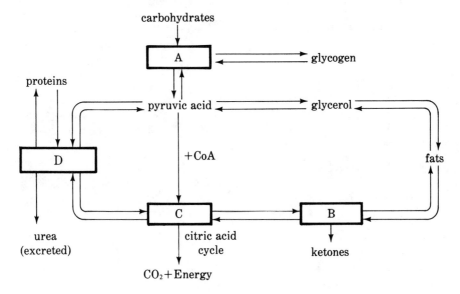

RECAP SECTION

Chapter 33 discusses the metabolism of carbohydrates and introduces many important general concepts of biochemistry. The roles of hormones and vitamins are of special interest for normal growth and development. A knowledge of metabolism is an integral part of many biological careers.

ANSWERS TO QUESTIONS

1. (1) saliva   (2) amylase   (3) stomach glands   (4) gastric juice
   (5) pancreas   (6) liver   (7) bile   (8) intestinal juice

2. (1) c   (2) d   (3) a   (4) b

3. (1) water
   (2) ATP
   (3) glycogen
   (4) Embden-Meyerhof
   (5) oxygen
   (6) citric acid
   (7) hormones
   (8) endocrine
   (9) acetyl-CoA

4. (1) d, h   (2) b   (3) g   (4) a, f   (5) c, e

5. A-9, B-5, C-2, D-7

6. A-13, B-10, C-4, D-8

CHAPTER THIRTY-FOUR
Metabolism of Lipids and Proteins

SELF-EVALUATION SECTION

1. Fill in the blank space or circle the appropriate response.

   The biological oxidation of fatty acids has been investigated since the early 1900s. Based on experimentation, Franz Knoop suggested that the chain was shortened by (1) _____ atoms at a time. This suggestion has been verified by other scientists. The carbon chain fragments can then feed into the Krebs cycle and yield high-energy phosphate in the form of (2) _____.

   The digestion of proteins provides alpha amino acids, which are absorbed through the intestinal wall into the blood stream. Amino acids undergo a variety of reactions while maintaining an equilibrium with the amino acid pool. When the amount of nitrogen excreted equals the nitrogen intake, a person is said to be in (3) _____. A growing person should take in more than is excreted, and this is termed a (4) _____ nitrogen balance. On the other hand, an individual who is fasting or undergoing starvation will be in a (5) _____ nitrogen balance. Absorbed amino acids are usually incorporated into body protein. They also can lose the amino group to become an alpha keto acid or undergo transamination. Transamination is an important reaction linking amino acids to the metabolic pathways of carbohydrates and lipids.

2. List the possible metabolic fates of amino acids in humans.

3. Complete the following reactions of amino acids. Name the amino acid reactant in both reactions.

(1) 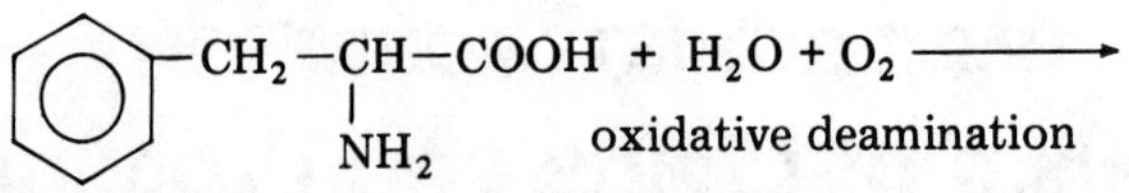 C₆H₅–CH₂–CH(NH₂)–COOH + H₂O + O₂ $\xrightarrow{\text{oxidative deamination}}$

(2) CH₂(NH₂)–COOH + COOH–CH₂CH₂–C(=O)–COOH $\xrightarrow{\text{transaminase}}$

4. Identify the terms or phrases given as referring to lipogenesis (L) or fatty acid oxidation (O).

   (1) catabolic pathway
   (2) cytoplasm
   (3) malonyl CoA plays a role
   (4) anabolic pathway
   (5) mitochondria
   (6) fatty acid chain linked to ACP
   (7) fatty acid chain linked to CoA

5. Refer to chapter 28, Table 28.1. How many acetyl CoA molecules can be produced by oxidation of the following fatty acids?

   (1) linolenic
   (2) stearic
   (3) myristic
   (4) capric
   (5) arachidic
   (6) caproic

6. Match the name of the biochemical process with the simplified reaction equation given.

   (1) 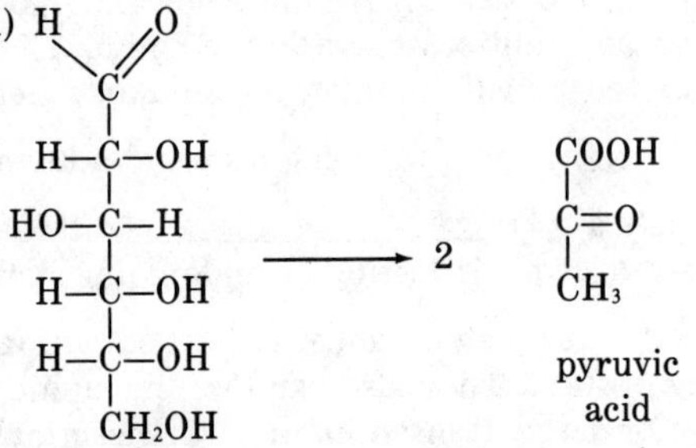 D-glucose → 2 pyruvic acid

   (2) phosphoenolpyruvate and ADP ⟶ ATP + pyruvate
   (3) Acetyl-CoA $\xrightarrow{[O]}$ CO₂ + H₂O + Energy
   (4) ADP + Pi + NADH + FADH₂ $\xrightarrow{[O]}$ H₂O + ATP
   (5) 6 CO₂ + 6 H₂O $\xrightarrow{\text{light}}$ C₆H₁₂O₆ + 6 O₂

   Process Names
   a. substrate level phosphorylation
   b. electron transport-oxidative phosphorylation
   c. photosynthesis
   d. Embden-Meyerhof pathway
   e. citric acid cycle

7. Which of the following items draw amino acids away from the amino acid pool? Which items contribute amino acids to the pool? Are any of the items two way processes?

   | amino acid pool |

   (1) amino acids from transamination
   (2) food
   (3) excreted nitrogen
   (4) non-protein nitrogen compounds
   (5) tissue protein

8. Fill in the blank space or circle the appropriate response.

   ATP (1) is/is not produced directly during β-oxidation of fatty acids but is formed when the reduced coenzymes (2) _____ and (3) _____ are oxidized by the mitochondrial electron system in concert with (4) _____ _____. Fatty acid oxidation is (5) aerobic/ anaerobic because $FADH_2$ and NADH can only be oxidized when (6) _____ is present.

RECAP SECTION

Chapter 34 is the final chapter in the discussion of biochemistry. It is appropriate that the last two chapters deal with the topic of metabolism of the three major sources of chemical energy. All nonphotosynthetic organisms, including humans, are dependent on the green plants for food. There are many remarkable similarities in the metabolic pathways of various organisms and the manner in which we all derive our energy from the food we consume. In fact, much of what we know about human nutrition and biochemistry was worked out with experiments involving organisms as simple as bacteria or as complex as rats and mice.

ANSWERS TO QUESTIONS

1. (1) two carbon (2) ATP (3) nitrogen balance (4) positive (5) negative

2. Incorporated into protein, utilized in other syntheses, deaminated to keto acid.

3. (1) phenylalanine + H_2O + O_2 ⟶ ⟨C$_6$H$_5$⟩—CH_2—$\overset{O}{\overset{\|}{C}}$—COOH + NH_3 + H_2O_2

   (2) glycine + HOOC—$CH_2CH_2\overset{O}{\overset{\|}{C}}$—COOH ⟶

   $\overset{O}{\overset{\|}{HC}}$—COOH + HOOC—$CH_2CH_2$—CH—COOH
   |
   NH_2

4. (1) O (2) L (3) L (4) L (5) O (6) L (7) O

5. (1) 9 (2) 9 (3) 7 (4) 5 (5) 10 (6) 3

6. (1) d. Embden-Meyerhof
   (2) a. substrate level phosphorylation
   (3) e. citric acid cycle
   (4) b. electron transport-oxidative phosphorylation
   (5) c. photosynthesis

7. amino acids away from pool (3), (4), (5)
   amino acids to the pool (1), (2), (5)
   two way processes (5)

8. (1) is not (2) $FADH_2$ (3) NADH (4) oxidative phosphorylation
   (5) aerobic (6) oxygen

CROSSWORD PUZZLE #2

Across

1. Graphite and diamond are two _____ forms of carbon.
2. Hydrocarbon with only single covalent bonds.
3. Diffusion of water through a semipermeable membrane.
4. The number of molecules in a mole is called _____ number.
5. Chemical symbol for element number 88.
6. A reaction in which an atom loses electrons.
7. An abbreviation for oxidation-reduction reactions.
8. Saccharides with two to six units linked together are _____ saccharides.
9. Diastereomers differing only at carbon number 1.

Down

1. The nucleic acid found outside the nucleus of a cell.
2. The force that a mass exerts on an area.
3. A substance that influences the rate of a reaction, but not the equilibrium of a reaction.
4. Because of the unequal sharing of covalent bond electrons, a water molecule is said to be _____.
5. The salt of a long-chain fatty acid is called a _____.
6. The pressure exerted by a gas in equilibrium with its liquid is called _____ pressure.
7. Torricelli (did/did not) invent the barometer.
8. Esters of high-molecular-weight fatty acids and high-molecular-weight alcohols.
9. Chemical symbol for element with atomic weight 131.3.
10. According to the Brønsted-Lowry Theory, an acid is a proton _____.
11. The nucleic acid found in the cell nucleus.
12. An electrically charged atom or group of atoms.
13. Chemical symbol for element number 31.

Answers are found at the end of Chapter 34.

ANSWERS
WORD SEARCH 1

H	C	O	M	P	O	U	N	D	D	I	S	O	T	O	P	E	S
O	X	C	S	Z	K	L	O	Y	P	M	N	Q	R	B	L	C	T
M	P	R	U	D	K	Q	T	V	F	V	G	M	W	E	Z	T	Y
O	I	Z	E	O	R	V	O	M	T	S	E	O	C	U	A	L	G
G	U	A	L	T	E	I	R	O	R	E	S	T	T	L	A	C	I
E	I	F	C	S	U	P	P	W	L	L	R	A	B	A	I	T	U
N	G	O	U	Q	E	G	Y	O	H	O	J	Y	G	R	N	O	P
E	E	E	N	E	U	T	R	O	N	P	A	M	V	E	E	S	E
O	L	B	U	I	S	W	H	E	J	I	M	N	M	T	K	O	R
U	S	O	V	T	Z	U	G	J	F	D	V	E	C	L	D	E	I
S	I	E	I	R	K	A	O	A	B	K	L	T	U	M	L	E	O
K	N	G	S	D	T	M	T	H	V	E	P	R	S	D	D	A	D
L	L	A	T	I	B	R	O	I	P	A	D	O	I	M	E	R	I
O	E	W	V	S	L	H	F	L	O	R	G	E	U	Q	N	V	C
B	O	I	O	N	N	A	P	A	E	N	O	L	T	W	S	Z	L
M	T	V	M	E	O	S	I	D	N	C	E	M	N	O	I	N	A
Y	L	S	A	T	R	T	Z	M	A	B	U	N	A	L	T	U	W
S	P	J	D	A	T	C	I	O	T	M	J	L	E	K	Y	V	Z
A	E	O	R	B	C	M	R	P	E	B	U	M	E	R	E	I	D
B	H	T	L	C	E	A	H	T	M	X	Y	U	T	T	G	G	C
G	D	I	U	F	L	O	A	O	F	E	S	N	X	G	A	Y	V
S	T	N	E	M	E	L	E	N	O	I	T	I	S	N	A	R	T
A	M	G	W	U	P	Q	N	S	W	D	L	E	R	V	K	C	B
H	I	V	R	E	T	E	M	O	R	D	Y	H	R	W	E	J	F

ANSWERS
CROSSWORD PUZZLE 1

								S		
H	E	T	E	R	O	G	E	N	E	O	U	S
	L				A				L	
	E		N		S	U	L	F	U	R
	M	E	T	A	L				T		X
P	E	N							I	D	E
O	N	E			T	A		C			N
	T	R	A	N	S	I	T	I	O	N		O
		G				N	O		M			N
R	A	Y		A	M		M		P		
	N			O		B	O	N	D	
	I			L		N		U		
	O	S		C	A	T	I	O	N	
	N	E		R				D		

314

ANSWERS
WORD SEARCH 2

A	E	G	Y	T	F	E	H	Y	K	B	D	I	S	P	D	N	J
G	P	I	S	O	L	U	T	E	O	Q	F	Y	D	I	L	K	Z
X	T	D	N	O	A	I	S	Y	C	I	E	V	L	A	H	P	O
C	K	E	B	G	L	O	M	R	L	X	R	U	N	F	O	J	Z
U	M	T	V	A	E	V	Z	M	A	O	T	E	Q	B	Q	Y	M
S	F	A	M	H	C	R	E	D	I	E	R	E	F	F	U	B	Q
D	N	R	U	S	H	E	R	N	C	S	Q	T	U	W	M	J	P
C	O	U	I	J	A	L	B	P	T	L	C	R	C	G	H	Z	W
N	I	T	N	U	T	B	I	N	E	H	A	I	V	E	E	Y	I
X	T	A	O	W	E	I	C	T	O	L	R	T	B	A	L	O	Q
K	A	S	R	D	L	C	V	M	O	E	W	M	Z	L	V	E	N
V	R	G	D	L	I	S	E	M	T	B	L	N	E	C	E	B	L
O	T	F	Y	X	E	I	I	O	N	I	Z	A	T	I	O	N	R
Y	I	G	H	W	R	M	H	Z	A	X	U	F	A	K	T	T	F
I	T	S	V	A	X	P	U	N	O	I	T	U	L	O	S	G	J
J	P	L	Q	Z	M	U	I	R	B	I	L	I	U	Q	E	E	T
M	C	X	S	A	H	T	A	U	D	F	N	Y	P	R	B	R	W

ANSWERS
CROSSWORD PUZZLE 2